31st Winter Workshop on Nuclear Dynamics (WWND 2015)

AA002687

Journal of Physics: Conference Series Volume 636

Keyston, Colorado, USA
25 – 31 January 2015

ISBN: 978-1-5108-1195-9
ISSN: 1742-6588

Printed from e-media with permission by:

Curran Associates, Inc.
57 Morehouse Lane
Red Hook, NY 12571

CURRAN ASSOCIATES INC. proceedings .com

Some format issues inherent in the e-media version may also appear in this print version.

Printed by Curran Associates, Inc. (2015)

Institute of Physics
Dirac House, Temple Back
Bristol BS1 6BE UK

Phone: 44 1 17 929 7481
Fax: 44 1 17 920 0979

techtracking@iop.org

Additional copies of this publication are available from:

Curran Associates, Inc.
57 Morehouse Lane
Red Hook, NY 12571 USA
Phone: 845-758-0400
Fax: 845-758-2634
Email: curran@proceedings.com
Web: www.proceedings.com

TABLE OF CONTENTS

Journal of Physics: Conference Series **636** (2015) 011001 doi:10.1088/1742-6596/636/1/011001

Preface

These are the proceedings of the 31st Winter Workshop on Nuclear Dynamics, which was held in Keystone, Colorado, in January 2015. As in previous years, the unique character of this conference series has allowed us to bring together nuclear scientists with very different interests to discuss recent progress and scientific achievements.

Out of the 63 contributions at WWND 2015 we have selected these 18 representative manuscripts. The topics capture the range of theoretical and experimental advances in our field. On the experimental side we saw very exciting results from the RHIC beam energy scan program, trying to characterize the exact energy required for a hadronic system to transition to deconfined Quark Gluon Matter. At the achievable LHC energies the focus was on the comparison of the data from the p–p, p–Pb and Pb–Pb runs. On the theory side this system size dependence of the experimental measurements led to a detailed evaluation of the initial conditions as well as a profound discussion on how small a QGP system can be. These results were complemented by the most recent continuum extrapolated data from lattice in order to model the complete evolution of the relativistic heavy ion system.

These proceedings of the 31st Winter Workshop on Nuclear Dynamics again provide a snapshot of the status of the field. The articles, many of which were written by some of the most promising young scientists in the field, are documenting the excitement and achievements that are characteristic for modern day nuclear science.

Rene Bellwied (University of Houston)
Frank Geurts (Rice University)
Anthony Timmins (University of Houston)

Journal of Physics: Conference Series **636** (2015) 011002 doi:10.1088/1742-6596/636/1/011002

Peer review statement

All papers published in this volume of *Journal of Physics: Conference Series* have been peer reviewed through processes administered by the proceedings Editors. Reviews were conducted by expert referees to the professional and scientific standards expected of a proceedings journal published by IOP Publishing.

31st Winter Workshop on Nuclear Dynamics (WWND2015) IOP Publishing
Journal of Physics: Conference Series **636** (2015) 012001 doi:10.1088/1742-6596/636/1/012001

Directed Flow of Charged Kaons in Au+Au Collisions from the BES Program at RHIC

Yadav Pandit (for the STAR Collaboration)
Department of Physics, UIC, Chicago, USA
E-mail: ypandit@uic.edu

Abstract.
We report the measurement of the directed flow (v_1) for charged kaons in Au+Au collisions at $\sqrt{s_{NN}}$ =7.7, 11.5, 19.6, 27, 39, 62.4 and 200 GeV as a function of rapidity and compare these results for pions, protons and antiprotons. These new kaon results may help to constrain the medium properties and collision dynamics including the in-medium kaon potential and baryon number transport in these collisions.

1. Introduction

The study of collective flow in relativistic nuclear collisions has potential to offer insight into the equation of state of the produced matter. Directed flow, v_1, is the first harmonic coefficient of the Fourier expansion of the final-state momentum-space azimuthal distribution [1],

$$v_1 = \langle \cos(\phi - \Psi_R) \rangle, \quad (1)$$

where the angle brackets indicate an average over all the particles and events used, ϕ denotes

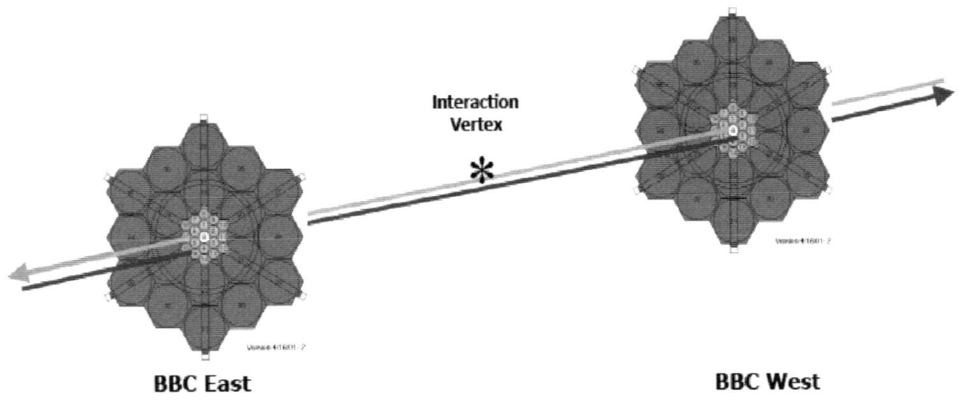

Figure 1. Beam Beam Counters at the STAR detector

the azimuthal angle of the outgoing particles and Ψ_R is the orientation of the event plane which is determined event-by-event. Both hydrodynamic [2] and nuclear transport models [3] indicate that directed flow is a promising observable for investigating a possible phase transition, especially in the region of Beam Energy Scan (BES) program at RHIC. In particular, the shape of v_1 as a function of rapidity, y, in the mid-rapidity region is of interest because it has been argued that it is sensitive to crucial details of the expansion of the participant matter during the early stages of the collision. The models indicate that the evolving shape and orientation of the participant zone and its surface play a role in determining the azimuthal anisotropy measured among these particles in the final state [4, 5].

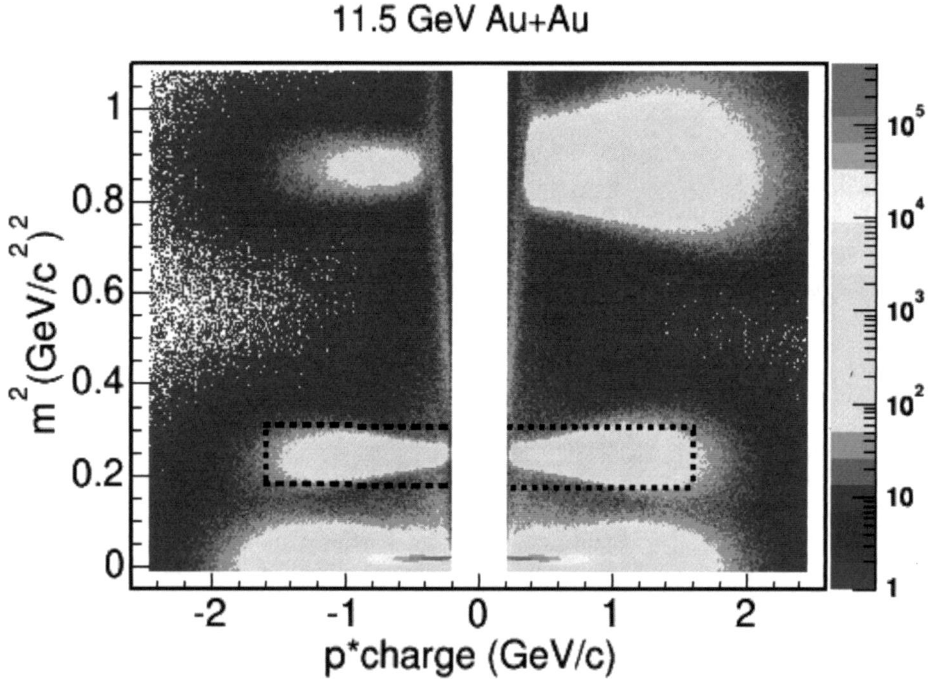

Figure 2. Mass squared based on time-of-flight information, versus momentum from TPC curvature, for the case of Au+Au collisions at 11.5 GeV. K^+ and K^- candidates are indicated by dashed-line boxes.

It is generally believed that the directed flow is generated early in the heavy-ion collision. Therefore it may be an important probe to provide valuable information of the onset of bulk collective dynamics during thermalization, even on the pre-equilibrium stage. Recently STAR published the measurement of directed flow of identified hadrons (protons, antiprotons, and positive and negative pions) with high precision for semi-central Au+Au collisions in the energy range $\sqrt{s_{NN}} = 7.7 - 200$ GeV [6]. After the publication of these data, there have been some theoretical efforts [7, 8] to understand the underlying physics and this is an ongoing effort. Still sizable discrepancies with experimental measurements in the directed flow characteristics are found for the microscopic kinetic models at $\sqrt{s_{NN}} < 20$ GeV and are common for the transport models HSD, PHSD and UrQMD. There are few other observables where kinetic approaches have a problem in this energy range [8].

The role of transported baryon number and absorption effects are not well understood. There

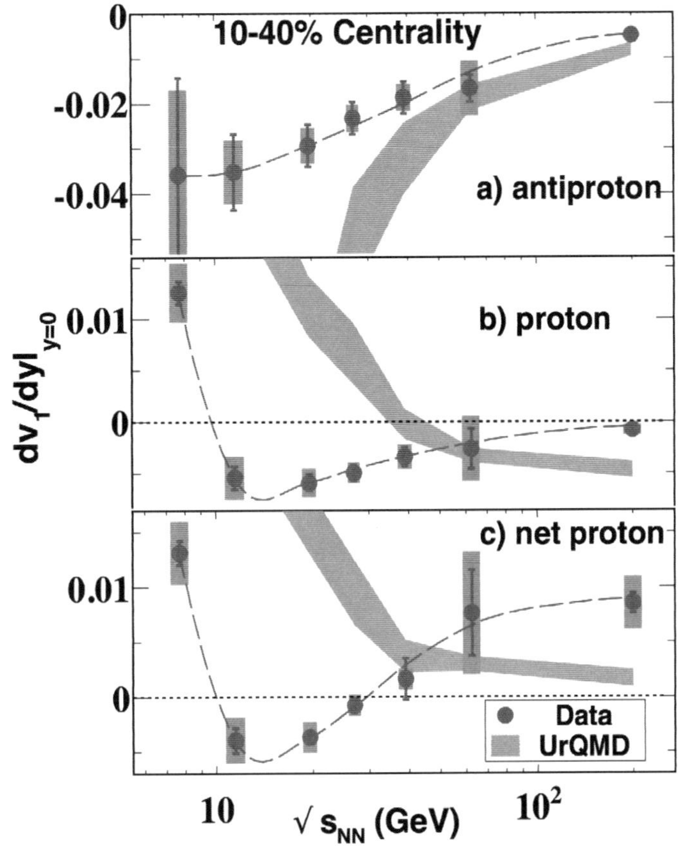

Figure 3. Directed flow slope (dv_1/dy) near mid-rapidity versus beam energy for intermediate-centrality Au+Au collisions. The slopes for protons, antiprotons and net-protons are reported [6].

are still a lot to be done, specially in the model calculation since the effect of event plane used in the experiment does not exactly coincide with the known reaction plane in the theory calculation. The role of produced particle vs spectator in the event plane calculation may change with energy since the event plane detector is in fixed phase space. UrQMD simulation suggests that these effects will have a minimal effect in the experimental results.

We report here the directed flow v_1 measurements for charged kaons in Au+Au collisions from the STAR experiment at the Relativistic Heavy Ion Collider (RHIC) in the range $\sqrt{s_{NN}}$ = 7.7 to 200 GeV, based on data from STAR experiment collected during RHIC run in the years 2010 and 2011. These measurements are presented as a function of rapidity and centrality.

2. The STAR experiment and Analysis Details

Data reported in these proceedings were collected in Au+Au collisions at $\sqrt{s_{NN}}$ = 7.7- 200 GeV in the year 2010 and 2011 with a minimum bias trigger. The main detector used in the present measurements is the Time Projection Chamber (TPC) [9], the primary tracking device at STAR, time of flight detector and beam beam counter . TPC has full azimuthal coverage and uniform acceptance in ±1.0 units of pseudorapidity. The charged particle momenta are measured by reconstructing their trajectories through the TPC. The centrality was determined

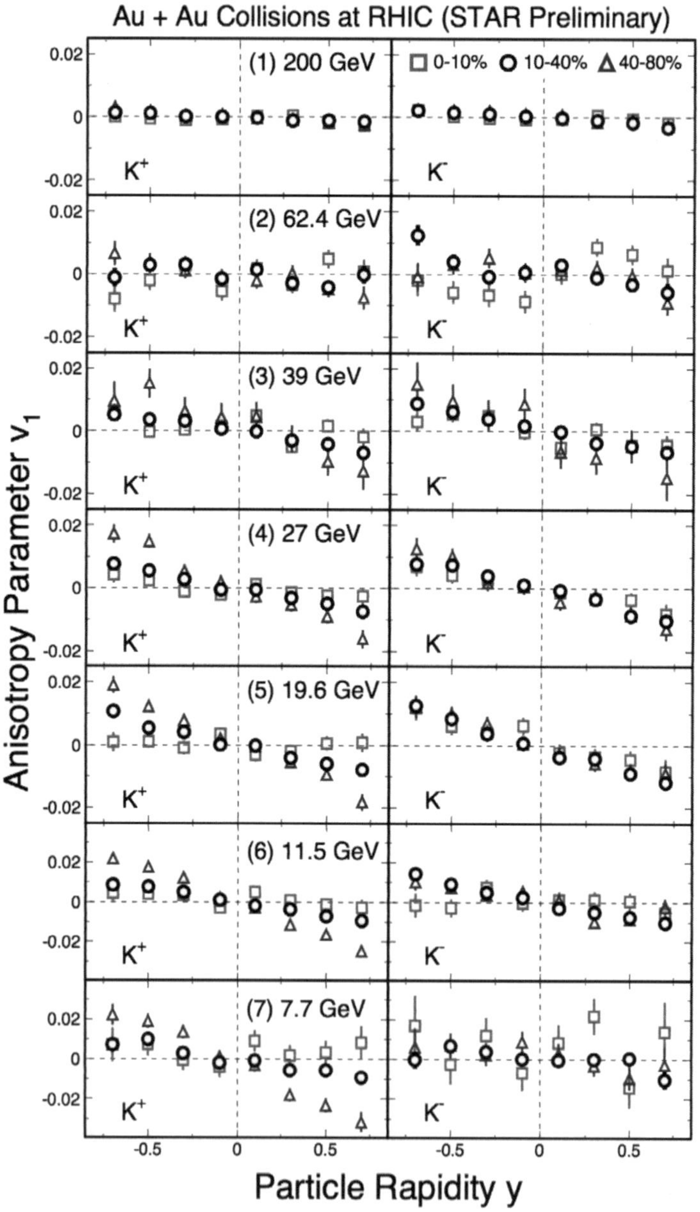

Figure 4. Directed flow for charged kaons K^+ on left panel and K^- on right panel versus rapidity for central (0-10%), intermediate-centrality (10-40%) and peripheral (40-80%) Au+Au collisions at $\sqrt{s_{NN}}$= 7.7-200 GeV. The plotted errors are statistical only, and systematic uncertainties are explained in the text.

by the number of tracks from the pseudorapidity region $|\eta| < 0.5$. The Beam Beam Counter (BBC) detector subsystem [10] consists of two detectors mounted around the beam pipe, each located outside the STAR magnet pole-tip at opposite ends of the detector approximately 375 cm from the center of the nominal IR as shown in Fig 1. Each BBC detector consists of nearly circular scintillator tiles arranged in four concentric rings that provide full azimuthal coverage. The two inner rings have a pseudorapidity coverage of $3.3 < |\eta| < 5.0$ and are used to reconstruct the first-order event plane for this analysis. Charged kaons were identified up to 1.6 GeV/c in transverse momentum based on specific energy loss in the TPC and from the time measurement by STAR's time-of-flight barrel in combination with the momentum.

Possible systematic uncertainties arising from non-flow, i.e., azimuthal correlations not related to the reaction plane orientation (arising from resonances, jets, strings, quantum statistics, and final-state interactions like Coulomb effects) are reduced due to the relatively large pseudorapidity gap between the STAR TPC and the BBC detectors [1]. Directed flow measurements based on the BBC event plane, where the BBC east and west detectors ensure symmetry in rapidity acceptance, cancel biases from conservation of momentum in the basic correlation measurement because the difference in v_1 between the rapidity hemispheres is used. However, momentum conservation effects do contribute to systematic uncertainty in the event-plane resolution, and thereby in the resolution-corrected signal, at the level of less than 2% [6]. The systematic uncertainty arising from particle misidentification and detector inefficiency is estimated by varying event and track cuts, and is $\sim$ 5%.

3. Result and discussion

In Fig. 4, directed flow for charged kaons, K^+ on left panel and K^- on right panel versus rapidity for central (0-10%), intermediate-centrality (10-40%) and peripheral (40-80%) Au+Au collisions at $\sqrt{s_{NN}}$= 7.7-200 GeV are shown. All data points in Fig. 4 are antisymmetric about mid-rapidity, verifying cancellation of the momentum conservation effect[11]. In intermediate and peripheral collisions, slopes of $v_1(y)$ near mid-rapidity for $K^\pm$ are generally negative for all energies with larger errors. Furthermore, STAR has previously reported negative slopes at mid-rapidity for pions for the energy range reported here and for protons the slopes were negative for 11.5 and higher beam energies [6].

Fig. 5 reveals the beam energy dependence of the directed flow slope dv_1/dy for $K^\pm$ near midrapidity for intermediate centrality (10-40%) along with protons, antiprotons and $\pi^\pm$. On the left panel we show results for K^+ compared with π^- and protons. On the right panel we compare K^- with π^+ and antiproton. New kaon measurements are shown as filled triangles, while results for protons, antiprotons and pions were previously reported in Ref. [6]. All slopes are based on a linear fit. Statistical and systematic errors are represented by bars and caps, respectively. Calculations from Hadron String Dynamics (HSD) [8] and UrQMD [7] are shown as shaded bands. At the lowest collision energy, $\sqrt{s_{NN}}$ = 7.7 GeV, charged kaons and all other produced hadrons show opposite sign of the mid-rapidity slope parameter compared to protons. In this low energy region, K^- v_1 slope is closer to zero than that of K^+. This observation supports the expectation that K^+ experiences a repulsive and K^- experiences attractive nuclear potential at these energies where baryon density is high. An extrapolation of STAR's systematic energy dependence is consistent with fixed-target results from the AGS. At the higher energy region, $\sqrt{s_{NN}} > 30$ GeV, where pair production becomes dominant, all observed particles seem to show a similar beam energy dependence, and the difference among them reduces as the energy increases.

4. Summary

In these proceedings, STAR results for directed flow $v_1(y)$ and directed flow slope (dv_1/dy) for charged kaons as a function of center of mass colliding energy and collision centrality are

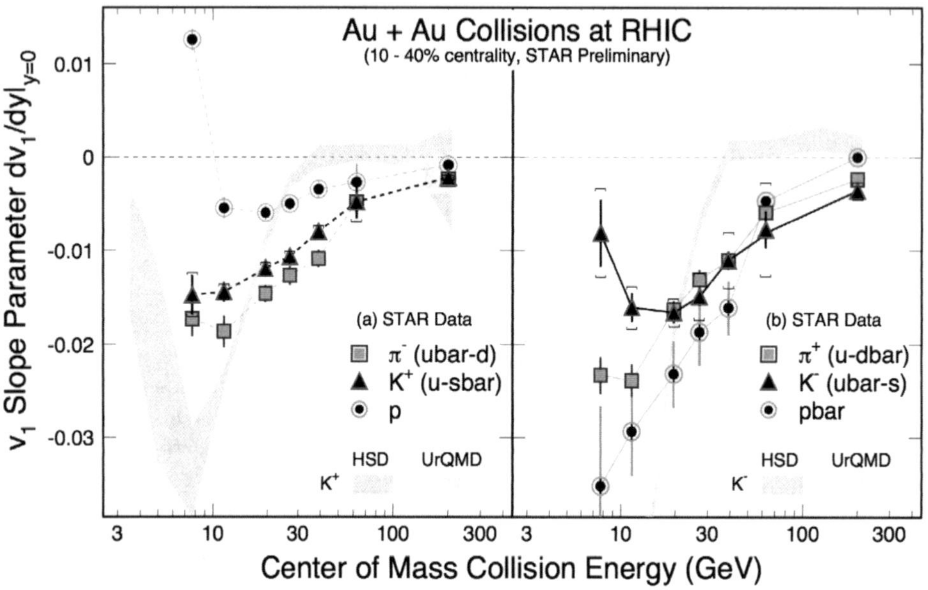

Figure 5. Beam energy dependence of the slope dv_1/dy at mid-rapidity for identified hadrons for intermediate centrality (10-40%). New kaon measurements are shown as filled triangles, while results for protons, antiprotons and pions were previously reported in Ref. [6]. All slopes are based on a linear fit. Statistical and systematic errors are represented by bars and caps, respectively. Calculations from Hadron String Dynamics (HSD) [8] and UrQMD [7] are shown as shaded bands.

presented. At the lowest collision energy, $\sqrt{s_{NN}} = 7.7$ GeV, charged kaons and all other produced hadrons show opposite sign of the mid-rapidity slope parameter compared with protons. At the higher energy region, $\sqrt{s_{NN}} > 30$ GeV, all observed particles seem to show a similar beam energy dependence, and the difference among them reduces as the energy increases. The transport model HSD, with the mean-field on, and UrQMD provide a reasonably good comparison to data at the high energy region, but fail at low energy. The present measurements might be helpful to further constrain the medium properties in terms of mean-field and the interplay of quark and baryon transport.

References

[1] A. M. Poskanzer and S. A. Voloshin, Phys. Rev. C **58**, 1671 (1998).
[2] U. W. Heinz, in *Relativistic Heavy Ion Physics*, Landolt-Boernstein New Series, Vol. I/23, ed. R. Stock (Springer Verlag, New York, 2010).
[3] S. A. Bass *et al.*, Prog. Part. Nucl. Phys. **41**, 225 (1998); M. Bleicher *et al.*, J. Phys. G **25**, 1859 (1999).
[4] H. Sorge, Phys. Rev. Lett. **78**, 2309 (1997).
[5] P. Kolb and U. Heinz, nucl-th/0305084; P. Huovinen and P.V. Ruuskanen, Annu. Rev. Nucl. Part. Sci. **56**, 163 (2006).
[6] L. Adamczyk *et al.* (STAR Collaboration), Phys. Rev. Lett. 112, 162301(2014)
[7] J. Steinheimer, J. Auvinen, H. Petersen, M. Bleicher and H. Stöcker, Phys. Rev. C 89, 054913(2014)
[8] V. P. Konchakovski, W. Cassing, Yu. B. Ivanov, and V. D. Toneev, Phys. Rev. C 90, 014903(2014)
[9] M. Anderson *et al.*, Nucl. Instr. Meth. A **499**, 659 (2003).

[10] C. A. Whitten (STAR Collaboration), AIP Conf. Proc. **980**, 390 (2008).
[11] N. Borghini, P. M. Dinh, J.-Y. Ollitrault, A. M. Poskanzer, S. A. Voloshin, Phys. Rev. C **66** 014901 (2002)

Heavy-flavour correlations in pp, p-Pb and Pb-Pb collisions with ALICE

S. Bjelogrlić (for the ALICE Collaboration)
Princetonplein 5, 3584 CC, Utrecht, the Netherlands
E-mail: sandro.bjelogrlic@cern.ch

Abstract. Heavy quarks (charm and beauty) are produced in initial hard scattering processes in heavy-ion collisions, before the formation of a strongly-interacting medium, the Quark-Gluon Plasma (QGP). The measurement of angular correlations between open heavy-flavour hadrons and charged particles can provide insight into the effects of the medium on charm and beauty production. For instance, in Pb-Pb collisions, the azimuthal correlations can provide information on the energy-loss mechanisms of heavy quarks in the QGP. Additionally, azimuthal correlations are sensitive to possible modifications of the heavy-quark parton shower and hadronisation in the presence of the medium. The observed double-ridge long-range correlations between light hadrons in p-Pb collisions could originate from a collective expansion of the system, as well as from gluon saturation in the initial state (color-glass condensate). The same effect can be studied for heavier quarks via the correlation between heavy-flavour hadrons (or their decay electrons) and charged particles. In pp collisions, the azimuthal correlations allow to measure the beauty production cross section as well as they represent a powerful tool to test pQCD models.

1. Introduction

The transverse-momentum (p_{T}) differential production cross section of D mesons and electrons from heavy-flavour hadron decays at central rapidity has been measured by the ALICE collaboration in pp, p-Pb and Pb-Pb collisions at $\sqrt{s}$ = 2.76 and 7 TeV, $\sqrt{s_{NN}}$ = 5.02 TeV and$\sqrt{s_{NN}}$ = 2.76 TeV, respectively [1, 2, 3, 4, 5, 6, 8, 7]. A strong suppression of the open-charm hadron yields was observed for $p_{\mathrm{T}} > 3$ GeV/c in Pb-Pb collisions. In p-Pb collisions no modification of the spectra was observed in the p_{T} range $1 < p_{\mathrm{T}}(\mathrm{D}) < 24$ GeV/c within the 15-20% uncertainties. This suggests that the suppression measured in Pb-Pb collisions can be attributed to a substantial in-medium energy loss of the heavy quarks [5].
Further insight into the charm energy loss mechanism in the QGP can be obtained with measurements of angular correlations between heavy-flavour hadrons (or their decay electrons) and the charged particles produced in the same di-jet as the heavy quark. The angular correlation distribution on the near side (i.e. the same azimuthal direction as the heavy-flavour "trigger" particle) is sensitive to possible modifications of the jet properties (e.g. shape and number of constituent particles) in the medium. On the other hand, the away side (i.e. the direction opposite in azimuth with respect to the trigger particle) is sensitive to the path-length dependence of the energy loss. The main observable to quantify the angular correlation

distribution is I_{AA}, defined as

$$I_{AA} = \frac{Y_{AA}}{Y_{pp}}, \quad \text{with } Y = \frac{1}{N_{trig}} \int \frac{dN^{assoc}(\Delta\varphi)}{d\Delta\varphi} d\Delta\varphi \tag{1}$$

which measures the associated yield in a given $\Delta\varphi$ (the difference in the azimuthal angle between the trigger and the associated particle) region and which can be compared in different collision systems.

In hadron-hadron correlations, measured in p-Pb collisions, a long-range azimuthal correlation, extended in pseudorapidity η, has been observed, after the jet contribution has been subtracted. This can be interpreted both as a final-state effect related to collective motion described by hydrodynamics or as an initial-state color-glass condensate (CGC) effect [9, 10]. One of the goals of the measurements of heavy-flavour triggered correlations in p-Pb collisions is to verify if this effect is observed also in the heavy-flavour sector.

In pp collisions, the azimuthal correlation of electrons from heavy-flavour decays with charged particles can be used to separate the beauty and charm contributions. This allows for the measurement of the relative fraction of electrons originating from a decay of a beauty hadron in the inclusive sample of electrons from heavy-flavour decays, that is then used to calculate the beauty cross section.

2. Heavy-flavour angular correlation analysis

Heavy-flavour production is measured in two different ways with ALICE at mid rapidity: by fully reconstructing D-meson hadronic decays and by measuring electrons originating from heavy-flavour hadron decays. D mesons are reconstructed in the ALICE central barrel via their hadronic decay channels, $D^0 \to K^-\pi^+$, $D^+ \to K^-\pi^+\pi^+$ and $D^{*+} \to D^0\pi^+ \to K^-\pi^+\pi^+$. The identification of the D mesons is possible thanks to the resolution of a few tens of μm on the track position close to the primary vertex provided by the Inner Tracking System (ITS). Indeed, the D-meson decay vertices are displaced from the primary vertex, and this feature is used to reduce the otherwise overwhelming combinatorial background. The measurement of the specific energy loss dE/dx with the Time Projection Chamber (TPC) and of the time of flight with the Time Of Flight (TOF) detector allows for a reliable identification of kaons and pions up to $p_T = 2$ GeV/c which is used to further suppress the combinatorial background. The invariant mass distribution of D-meson candidates is fitted with a gaussian distribution for the signal and a parametrisation for the background. The D-meson candidates, whose reconstructed invariant mass falls within a $\pm 2\ \sigma$ (the width of the Gaussian function obtained from the fitting procedure) window around the nominal D-meson mass are paired with charged particles produced in the collision, and for each "trigger"-associated particle pair the difference in pseudorapidity ($\Delta\eta$) and azimuthal angle ($\Delta\varphi$) is calculated. Every correlation pair in the obtained two dimensional ($\Delta\varphi$,$\Delta\eta$) distribution is weighted with the inverse of the trigger particle reconstruction efficiency and the inverse of the reconstruction efficiency of the associated tracks. The resulting angular correlation distributions include the contributions of both signal and background candidates. The latter contribution is subtracted using the sidebands (4-8 σ region) of the D-meson peak in the invariant mass distribution to build the background distribution.

Electrons are identified using the TPC and TOF detectors and by exploiting the measurement of the electron energy E with the Electromagnetic Calorimeter (EMCal). The inclusive sample of electrons does not only contain electrons from heavy-flavour decays, but also photonic electrons. Photonic electrons are electrons from Dalitz decays of light mesons (in particular π^0 and η) and electrons from photon conversions, that are identified using the invariant mass of the electron-positron pair[14, 6]. The ($\Delta\varphi$,$\Delta\eta$) distribution is calculated in the same way as for the D meson, and the respective reconstruction efficiency corrections are applied.

To correct for the effects of detector inhomogenities and of limited detector acceptance, the event mixing technique is used. D mesons (or electrons) are paired with charged tracks from other events with similar multiplicity and position of the collision vertex along the beam axis. The obtained distribution is normalized to 1 for $(\Delta\varphi,\Delta\eta) = (0,0)$ and the inverse of the latter is used as a weight map for the distribution obtained from correlations in the same event.
In addition, for the D meson–charged particle azimuthal correlation analysis, the angular distribution of D mesons originating from B-meson decays must be subtracted. The fraction of the latter is computed using the efficiency calculated for prompt and secondary D mesons and the p_{T}-differential cross sections obtained from a FONLL perturbative QCD calculation [1, 15], while the correlation distribution of D mesons from B decays with charged particles is obtained from simulations performed with the PYTHIA generator [11].
The fully corrected distribution, obtained after normalization to the number of heavy-flavour candidates, is studied differentially as function of the transverse momentum of the trigger particle and of the associated tracks and, in the case of the electrons, where the available statistics allows this, also as function of the event multiplicity (p-Pb) and centrality (Pb-Pb).

3. Results

The D meson–charged particle azimuthal correlation analysis has been performed on a data sample of $3.14 \cdot 10^8$ minimum bias pp collisions and $1.3 \cdot 10^8$ minimum bias p-Pb collisions. The minimum-bias trigger required the arrival of bunches from both directions and coincident signals in both scintillator arrays of the V0 detector, covering the regions $2.8 < \eta < 5.1$ and $-3.7 < \eta < -1.7$. The analysis was performed in three D-meson p_{T} intervals in pp collisions ($3 < p_{\mathrm{T}}\,(\mathrm{D}) < 5$ GeV/c, $5 < p_{\mathrm{T}}\,(\mathrm{D}) < 8$ GeV/c and $8 < p_{\mathrm{T}}\,(\mathrm{D}) < 16$ GeV/c) and in two p_{T} intervals in p-Pb collisions ($5 < p_{\mathrm{T}}\,(\mathrm{D}) < 8$ GeV/c and $8 < p_{\mathrm{T}}\,(\mathrm{D}) < 16$ GeV/c). Different minimum thresholds for the associated track p_{T} were studied ($p_{\mathrm{T}}^{assoc} >$0.3, 0.5 and 1 GeV/c).

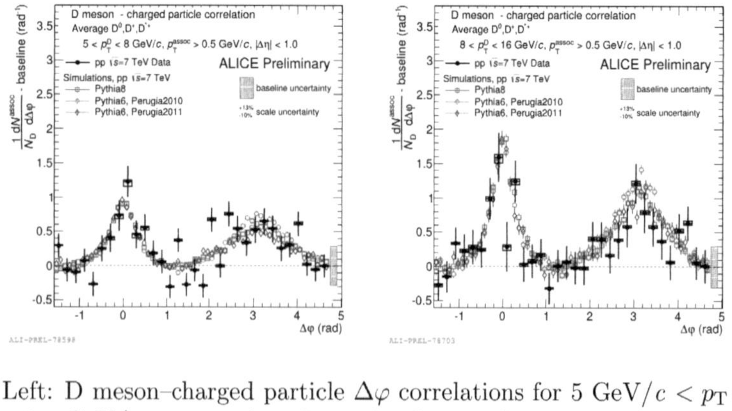

Figure 1. Left: D meson–charged particle $\Delta\varphi$ correlations for 5 GeV/$c < p_{\mathrm{T}}\,(\mathrm{D}) < 8$ GeV/c and $p_{\mathrm{T}}^{assoc} > 0.5$ GeV/c compared with results from different PYTHIA generators [11, 12]. Right: Same as left panel, but for 8 GeV/$c < p_{\mathrm{T}}\,(\mathrm{D}) < 16$ GeV/c.

Figure 1 shows the D meson–charged particle azimuthal correlation in pp collisions for 5 GeV/c $< p_{\mathrm{T}}\,(\mathrm{D}) < 8$ GeV/c (left panel) and 8 GeV/$c < p_{\mathrm{T}}\,(\mathrm{D}) < 16$ GeV/c (right panel), with $p_{\mathrm{T}}^{assoc} > 0.5$ GeV/c. The distributions from data are compared to the distributions obtained from Monte-Carlo simulations obtained with different tunes of the PYTHIA generator [11, 12].

Journal of Physics: Conference Series **636** (2015) 012002 doi:10.1088/1742-6596/636/1/012002

The baseline, that is due to the uncorrelated background, is calculated from the transverse region $\pi/4 < |\Delta\varphi| < \pi/2$, and has been subtracted in order to study the near-side and away-side peak properties. The variation of the baseline upon redefinition of the above mentioned interval is treated as a systematic uncertainty. All the different Monte-Carlo distributions describe well the distributions from data within the relatively large statistical uncertainties.

The left panel of Fig.2 shows the azimuthal correlation distribution between D mesons and

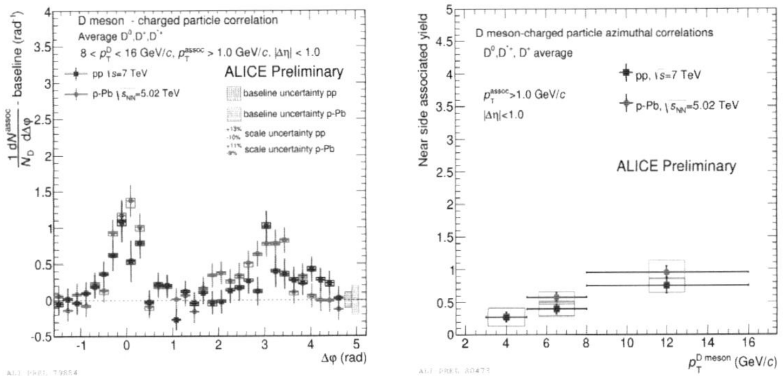

Figure 2. Left: D meson–charged particle $\Delta\varphi$ correlations for 8 GeV/c < $p_{\rm T}$ (D) < 16 GeV/c and $p_{\rm T}^{assoc}$ >1 GeV/c. Right: near-side yield as function of the D-meson $p_{\rm T}$.

charged particles in the interval $|\Delta\eta| < 1$ for $8 < p_{\rm T}$ (D) < 16 GeV/c and $p_{\rm T}^{assoc}$ >1 GeV/c. The azimuthal correlation distributions measured in pp and p-Pb collisions agree within uncertainties after subtracting the baseline due to the uncorrelated background. The same is observed also for the other minimum thresholds of the associated tracks that were studied. The azimuthal correlation distributions are fitted with a double Gaussian (one for the near-side and one for the away-side peaks). The near-side associated yields are obtained as the integral of the Gaussian distribution that is centered at $\Delta\varphi = 0$. The main source of systematic uncertainty is the definition of the baseline as described above, and it is estimated by recalculating the near side yield upon redefinition of the interval where the baseline is computed. The right panel of Fig. 2 shows the trends of the near-side associated yield as function of the D-meson $p_{\rm T}$ in pp and p-Pb collisions, confirming the agreement within uncertainties already observed in the comparison of the azimuthal correlation distributions in the two collision systems.

The angular correlation distribution between heavy-flavour decay electrons and charged particles has been measured in pp, p-Pb and in Pb-Pb collisions [7]. The pp analysis has been performed on the EMCal-triggered data sample at $\sqrt{s} = 2.76$ TeV, containing $6.4 \cdot 10^5$ events. The measured angular correlation of electrons from heavy-flavour decays and charged particles is composed of electrons originating from beauty hadrons and charm hadrons. By exploiting the different decay kinematics (that is due to the fact that beauty hadrons are heavier than charm hadrons) the two contributions can be separated, and the fraction of electrons from beauty decays, r_B, can be extracted using PYTHIA Monte-Carlo templates [11]. The fraction r_B is defined as

$$r_B = \frac{N_B}{N_C + N_B} \qquad (2)$$

where N_B represents the total number of electrons originating from beauty decays, while N_C represents the total number of electrons originating from charm decays.

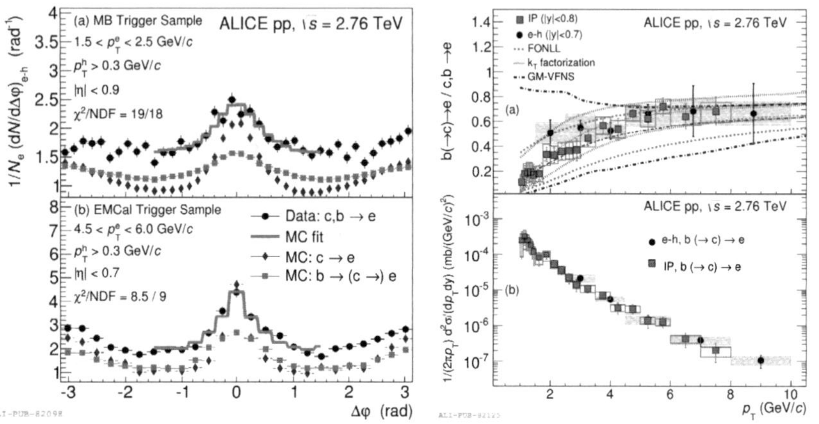

Figure 3. Upper left: azimuthal correlation of electrons from heavy-flavour decays and charged particles in pp collisions at $\sqrt{s} = 2.76$ TeV, for 1.5 GeV/c < $p_{\rm T}$ (e) < 2.5 GeV/c and $p_{\rm T}^{assoc}$ >0.3 GeV/c. Lower left: same as Upper left, but for 4.5 GeV/c < $p_{\rm T}$ (e) < 6 GeV/c. Upper right: r_B fractions using the azimuthal correlation method as function of the electron $p_{\rm T}$compared to the r_B ratio using the impact parameter method [13] and different model predictions. The dashed line represent different model prediction. Lower right: The beauty cross section as function of the beauty-decay electron $p_{\rm T}$, following the same colouring scheme as the upper-right figure.

The left panel of Fig.3 shows the heavy-flavour decay electron (HFE)-charged particle $\Delta\varphi$ distribution from data and those from the Monte-Carlo templates: $\Delta\varphi^{C}_{templ}$, the template of electrons from charm-hadron decays and, $\Delta\varphi^{B}_{templ}$, the template of electrons from beauty-hadron decays. These measured distributions are fitted following the parametrisation:

$$\Delta\varphi = const + r_B \Delta\varphi^{B}_{templ} + (1 - r_B)\Delta\varphi^{C}_{templ} \quad (3)$$

From the fitting procedure it is possible to extract r_B and the result is shown in the upper right panel of Fig. 3 as function of the electron $p_{\rm T}$. The r_B ratio obtained from azimuthal correlations is in good agreement with the one obtained using the impact parameter method [13] and the one obtained from FONLL predictions [15]. The lower right panel of Fig. 3 shows the beauty-decay electron production cross section, calculated by multiplying the inclusive heavy-flavour cross section with the r_B fraction.

The Pb-Pb analysis has been performed in two centrality classes: central (0-8%) and semi-central (20-50%) collisions, containing $1.6 \cdot 10^7$ and $9.5 \cdot 10^6$ events, respectively. The datasets have been acquired using centrality triggers based on the measured V0 amplitude. The left panel of Fig. 4 shows the comparison of the azimuthal correlations between heavy-flavour decay electrons and charged particles in pp collisions and in central and semi-central Pb-Pb collisions for 6 < $p_{\rm T}$ (e) < 8 GeV/c and 4 < $p_{\rm T}^{assoc}$ < 6 GeV/c. The baseline, including a modulation reflecting the correlation due to elliptic flow [14], has been subtracted from the Pb-Pb distributions. Within the statistical uncertainties, the three correlation distributions agree with each other. The right panel of Fig. 4 shows I_{AA} for the central and semi-central collisions. I_{AA} is compatible with one within the uncertainties. The current statistics does not allow to conclude on a possible modification of the near-side structure.

The electron–charged particle correlation distribution in p-Pb collision is shown in Fig. 5. The left panel shows a comparison of the correlation distributions for different p-Pb multiplicity classes from V0A (forward multiplicity detector with 2.8< η <5.1) and for pp collisions for 1

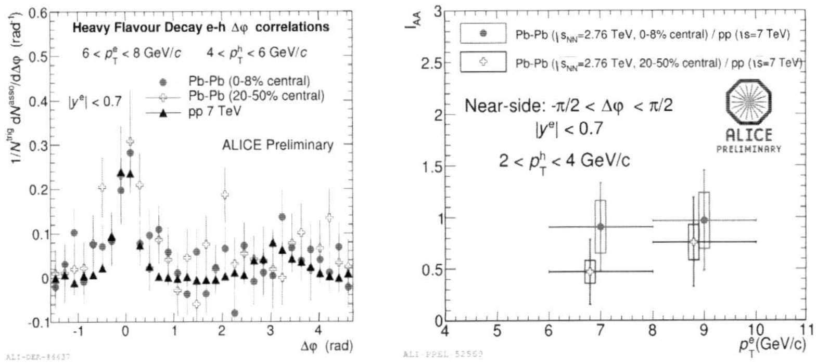

Figure 4. Left: Azimuthal correlations between heavy-flavour decay electrons and charged particles in pp collisions, central and semi-central Pb-Pb collisions for 6 GeV/c < $p_{\rm T}^{\rm e}$ < 8 GeV/c and 4 < $p_{\rm T}^{assoc}$ < 6 GeV/c. Right: near side I_{AA} as function of $p_{\rm T}$ (e).

< $p_{\rm T}$ (e) < 2 GeV/c and 0.5 < $p_{\rm T}^{assoc}$ < 2 GeV/c. A multiplicity dependence of the near-side and away-side correlation peaks is observed: the 60-100% p-Pb lowest multiplicity class is compatible with pp collisions, while an increase of the correlation yield is observed moving towards higher multiplicities.
After subtracting the correlation distribution of the 60-100% multiplicity class from the one of the 0-20% multiplicity class to remove jet-like contributions, within the limited statistics a long-range double-ridge structure is observed as shown in the right panel of Fig. 5. This distribution is qualitatively similar to the one observed in the hadron-hadron correlation analysis (dominated by light-flavour particles) [9]. This suggests that the mechanism that generates these long-range correlation structures may affect also heavy flavours.

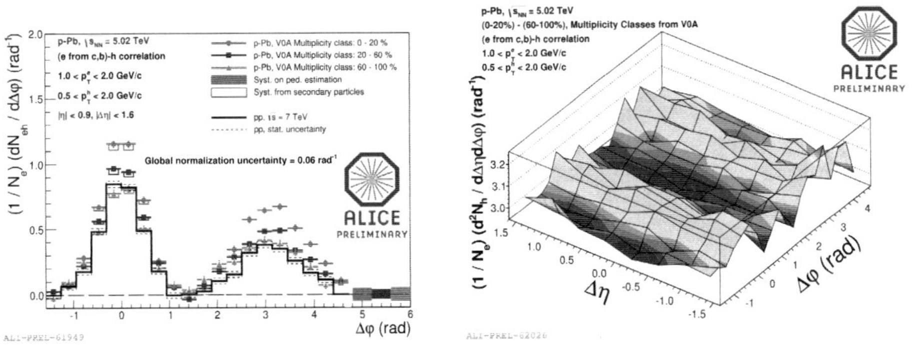

Figure 5. Left: Angular correlation between heavy-flavour decay electrons and charged particles in pp and p-Pb collisions with 1 GeV/c < $p_{\rm T}^{\rm e}$ < 2 GeV/c with 0.5 GeV/c < $p_{\rm T}^{assoc}$ <2 GeV/c for three different multiplicity intervals. Right: ($\Delta\varphi$,$\Delta\eta$) correlations in the 20% highest multiplicity p-Pb collisions after subtracting the correlations in the multiplicity class 60-100%.

4. Conclusions

ALICE has measured D meson–charged particle azimuthal correlations in pp and p-Pb collisions and correlations of electrons from heavy-flavour decays with charged particles in pp, p-Pb and Pb-Pb collisions. Given the current uncertainties, from the comparison of the D meson–charged particle azimuthal correlation distributions in pp and p-Pb collisions it is not possible to conclude on any possible modification due to cold nuclear matter effects. In pp collisions, the correlation distributions are in agreement within the uncertainties with the calculation from PYTHIA Monte-Carlo simulations. The fraction of electrons originating from beauty-hadron to those originating from all heavy-flavour hadron decays has been measured through the azimuthal correlations and it is compatible with the fraction measured using the impact parameter method and with the model predictions.

The measured correlation distributions for the heavy-flavour decay electron–charged particle azimuthal correlation in central and semi-central Pb-Pb collisions are compatible, within uncertainties, with the one measured in pp collisions. The measured near-side I_{AA} is compatible with unity. Due to limited statistics of the Pb-Pb data sample, the current results do not allow to conclude on possible modifications of the heavy-flavour jet structure induced by medium effects. In p-Pb collisions, the $(\Delta\varphi,\Delta\eta)$ correlation between heavy-flavour decay electrons and charged particles shows a long-range double-ridge structure resembling that observed in hadron-hadron correlations, suggesting that the same effect (hydrodynamics or CGC) might play a role also in the heavy-flavour sector. A reduction of the currently large statistical uncertainties is expected with the new data from LHC run II as well as from the ALICE ITS upgrades (after 2018).

[1] B. Abelev *et al.* (ALICE Collaboration) 2012, JHEP 01 128
[2] B. Abelev *et al.* (ALICE Collaboration) 2012, JHEP 09 112
[3] B. Abelev *et al.* (ALICE Collaboration) 2012, Phys.Rev.Lett. 113 (2014) 23, 232301
[4] B. Abelev *et al.* (ALICE Collaboration) 2015, Phys.Rev. D91 (2015) 012001
[5] A. Festanti for the ALICE Collaboration, Nucl.Phys. A931 (2014) 514-519
[6] R. Bailhache for the ALICE Collaboration, Nucl.Phys. A931 (2014) 530-534
[7] E. Pereira de Oliveira Filho for the ALICE collaboration, Nucl.Phys. A932 (2014) 258263
[8] S. Li for the ALICE Collaboration, Nucl.Phys. A931 (2014) 546-551
[9] B. Abelev *et al.* (ALICE Collaboration), Phys. Lett. B719 (2013) 29-41
[10] L. Milano for the ALICE collaboration, Nucl.Phys. A931 (2014) 1017-1021
[11] T. Sjostrand, S. Mrenna, P. Skands, 2006, arXiv:hep-ph/0603175
[12] P. Z. Skands, 2010, arXiv:1005.3457v5
[13] B. Abelev *et al.* (ALICE Collaboration), Phys. Let. B 738 (2014) 97108
[14] A. Dubla for the ALICE Collaboration, 2013, arXiv:1311.5429v2
[15] M. Cacciari, M. Greco and P. Nason, JHEP 9805 (1998) 007; M.Cacciari *et al.* JHEP 0103 (2001) 006

Journal of Physics: Conference Series **636** (2015) 012003 doi:10.1088/1742-6596/636/1/012003

Results on open-charm production in pp, p-Pb and Pb-Pb collisions with ALICE at the LHC

E. Meninno for the ALICE Collaboration

Via Giovanni Paolo II, 132 - 84084 - Fisciano (SA), Italy.

E-mail: elisa.meninno@cern.ch

Abstract. ALICE (A Large Ion Collider Experiment) is designed to study the strongly-interacting medium created in heavy-ion collisions at LHC energies, the Quark-Gluon Plasma (QGP). Charm and beauty quarks are powerful probes to study the QGP in heavy-ion collisions: produced in hard partonic scattering processes on a short time scale, they are expected to traverse the QCD medium, interacting with its constituents and losing energy through radiative and collisional processes. In ALICE, open-charm production is studied through the reconstruction of the hadronic decays of D^0, D^+, D^{*+} and D_s^+ mesons at mid-rapidity. The high precision tracking, good vertexing capabilities and excellent particle identification offered by ALICE allow for the measurement of particles containing heavy quarks (particularly D mesons) in a wide transverse-momentum range in pp, p-Pb and Pb-Pb collisions. A review of the main results on D-meson production in pp collisions at $\sqrt{s}$ = 7 TeV, Pb-Pb collisions at $\sqrt{s_{NN}}$ = 2.76 TeV and the most recent results in p-Pb collisions at $\sqrt{s_{NN}}$ = 5.02 TeV will be presented. In particular, the p_T-differential yields and cross sections in the three collision systems, the nuclear modification factors R_{AA} and R_{pPb} in Pb-Pb and p-Pb collisions, and the elliptic flow in Pb-Pb collisions will be discussed. The D-meson yield in pp and p-Pb collisions will also be shown as a function of charged-particle multiplicity.

1. Introduction

The ALICE experiment at the LHC aims at investigating the properties of the Quark-Gluon Plasma, the hot and dense state of strongly-interacting matter produced in high-energy heavy-ion collisions. Charm and beauty quarks are among to the most powerful probes to study the QGP in heavy-ion collisions: produced in hard parton scattering processes occurring in the early stages of the collision, they traverse the QCD medium, interact with its constituents and experience the whole evolution of the medium. Open heavy-flavour hadron production is expected to be sensitive to the energy density of the system through the mechanism of in-medium energy loss of heavy quarks [1-2-3]. Quantum Chromodynamics (QCD) theoretical calculations predict a dependence of the energy loss on the colour charge and on the mass of the parton traversing the medium. This results in an expected hierarchy of the parton energy loss, with beauty quarks losing less energy than charm quarks, and charm quarks losing less energy than light quarks and gluons. One of the observables that are sensitive to the interaction of hard partons with the medium is the nuclear modification factor, R_{AA}, defined as the ratio of the particle yield measured in Pb–Pb collisions and the cross section in pp collisions scaled with the nuclear overlap integral. R_{AA} is expected to be unity for heavy flavours in the absence of medium effects. The expected hierarchy of the energy loss described above can be probed comparing the R_{AA} of different particle species, namely B-, D- and light-hadron R_{AA}. Further knowledge of the properties of the medium created in heavy-ion collisions can be gained from the study of the azimuthal anisotropy of open heavy flavours: the initial spatial asymmetry is transformed into an asymmetry in momentum via hydrodynamic expansion of the medium. This is quantified in terms of the second coefficient v_2 in a Fourier expansion of the D-meson azimuthal distribution. The elliptic flow v_2 brings information on the medium transport properties: on the question whether heavy quarks take part in the collective expansion of the medium, and on the path-length dependence of energy loss. The interpretation of particle production measurements in Pb-Pb collisions requires detailed studies and understanding of their production in pp and p-Pb collisions: pp collisions provide the essential reference for the nuclear modification factor and a sensitive test of perturbative QCD models describing the production of heavy flavours in elementary hadronic collisions at LHC energies. The study of p–Pb collisions is crucial to access cold nuclear matter (CNM) effects in the initial and final state (modification of the Parton Distribution Functions in nuclei nPDF [4-5], gluon saturation at low Bjorken-x [6], k_T broadening), assuming that an extended, long-lived QGP is not formed in these collisions.

2. Open heavy-flavour measurements with ALICE

The ALICE detector, described in detail in [7], consists of a central barrel at mid-rapidity, a muon spectrometer at forward rapidity and a set of detectors (forward and backward rapidities) for global collision characterization and triggering purposes.

Open heavy flavours are measured in several charm hadronic decay channels and via electrons from semi-leptonic decays of charm and/or beauty hadrons at mid-rapidity. At forward rapidity, open heavy-flavour production is studied in the semi-muonic decay channel. Electrons are identified at mid-rapidity through their specific energy loss in the TPC gas combined with the information from the Time-Of-Flight (TOF) detector and from the electromagnetic calorimeter (EMCal). Muons are reconstructed in the five tracking stations of the Muon Spectrometer ($-4 < \eta < -2.5$).

2.1. D-meson reconstruction and selection

D mesons are reconstructed in ALICE via their hadronic decays $D^0 \to K^-\pi^+$, $D^+ \to K^-\pi^+\pi^+$, $D^{*+} \to D^0\pi^+ \to K^-\pi^+\pi^-$, $D^+_s \to \phi\pi^+ \to K^-K^+\pi^+$ and their charge conjugates.

D-meson selection is based on the reconstruction of decay vertices displaced by a few hundred μm from the interaction vertex, exploiting the high track-position resolution close to the interaction vertex provided by the Inner Tracking System (ITS). The large combinatorial background is reduced by selections applied on the decay topology and by the identification of charged kaons and pions via their specific energy loss in the TPC and their time of flight measured with the TOF detector. Raw D-meson yields are obtained from an invariant mass analysis of the pair/triplet of candidates. Then, the contribution from B-meson decay feed-down is subtracted, in order to obtain the prompt D-meson yields.

3. Results

Cross section measurements of prompt D mesons were performed with ALICE in pp collisions at $\sqrt{s}$ = 7 TeV and 2.76 TeV [8-10] and are well described by perturbative QCD calculations. The p_T-differential cross sections for D^0 and D^{*+} at $\sqrt{s}$ = 7 TeV are shown in Fig.1, together with FONLL [11] and GM-VFNS [12] calculations. Within uncertainties, theoretical predictions and measurements agree with each other. Nevertheless, it can be noted that the measurements tend to be higher than the central value of the FONLL predictions, as it was observed at lower collision energies, at RHIC and at the Tevatron [13-14]. For GM-VFNS, instead, the data lie on the lower side of the predictions.

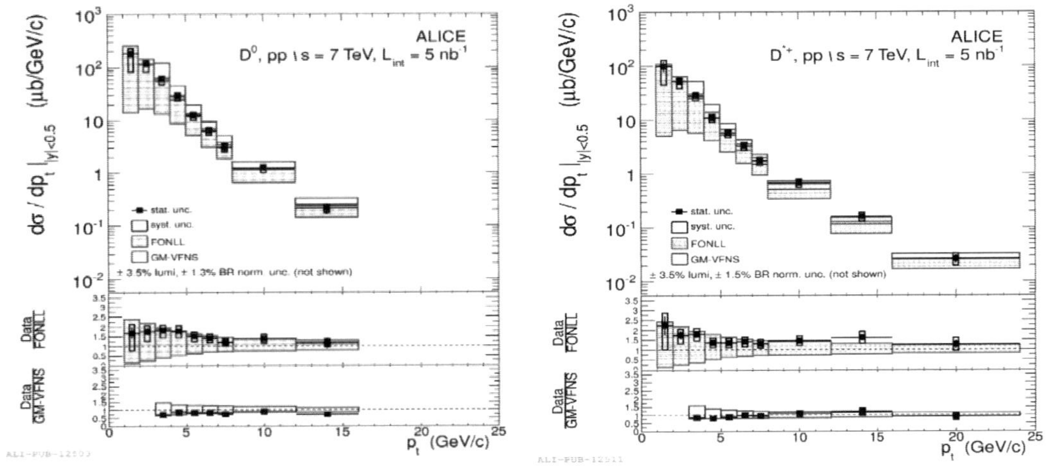

Figure 1: **Top**: p_T-differential cross section for prompt D^0 and D^{*+} mesons in pp collisions at $\sqrt{s}$= 7 TeV compared with FONLL and GM-VFNS theoretical predictions. **Bottom**: the ratio of the measured cross section and the central FONLL and GM-VFNS calculations.

In p–Pb collisions at $\sqrt{s_{NN}}$=5.02 TeV, CNM effects such as the modifications of the PDFs due to the presence of the nucleus are expected to affect the heavy-quark yield and p_T distributions relative to pp collisions. In particular, by measuring heavy-flavour hadron production in different p_T ranges it is possible to access different Bjorken-x regimes. The nuclear modification factor R_{pPb} quantifies the D-meson cross section in p–Pb collisions relative to the one in pp collisions scaled by the atomic mass number A = 208 of the lead nucleus. The average of the R_{pPb} for D^0, D^+ and D^{*+} in the p_T range $1 < p_T < 24$ GeV/c, is shown in Fig. 2 together with the comparison with theoretical calculations. It can be observed that the R_{pPb} of prompt D mesons is described within uncertainties by different models including initial-state effects [15]. The measurements confirm that initial- and final-state effects due to the presence of CNM are small in the measured p_T range. The measurements are well described by

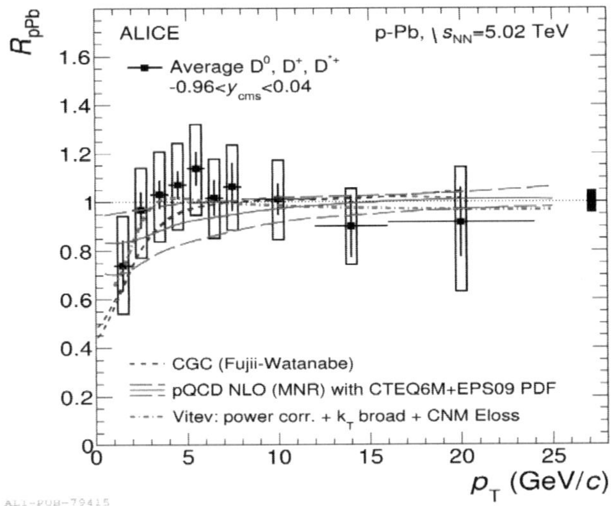

Figure 2: Average R_{pPb} as a function of p_T for prompt D^0, D^+ and D^{*+} mesons in p–Pb collisions at $\sqrt{s_{NN}}$ = 5.02 TeV, compared with models. Statistical (bars), systematic (empty boxes) and normalization (full box) uncertainties are shown.

theoretical predictions based on pQCD calculations including the EPS09 [16] nPDFs parametrisation, with calculations based on the Color Glass Condensate (CGC) and with a model including cold-nuclear matter energy loss, nuclear shadowing and k_T-broadening [17-18].
It has been observed that Multi-Parton Interactions, MPIs, multiple hard scatterings occurring between incoming partons in the same hadronic collision, play a relevant role in particle production at LHC energies. In particular, CMS results on jets and underlying event production rates and $\langle p_T \rangle$ as a function of the charged-particle multiplicity show a better agreement with models including MPI contribution [19]; measurements with ALICE of minijets, interpreted in the PYTHIA model, indicate an increase of MPIs with increasing multiplicity of charged particles in pp collisions [20]. Studies as a function of the event multiplicity were also performed by the ALICE Collaboration for prompt D mesons, inclusive J/Ψ in both pp and p-Pb collisions and for non-prompt J/Ψ in pp collisions. In Fig. 3 (left), average D^0, D^+, D^{*+} yield per-event in pp collisions, in multiplicity intervals normalized to the same yield obtained in the multiplicity-integrated sample is reported as a function of the charged-particle multiplicity normalized by its average measured in multiplicity-integrated events [21]. The results in the different p_T ranges shown in the figure are in agreement within the uncertainties. An increasing trend of the D-meson yield as a function of charged-particle multiplicity in pp collisions is observed, suggesting that MPIs, which substantially contribute to high-multiplicity events, are affecting heavy-flavour production. A similar trend is observed in p–Pb collisions, as shown in the right panel of Fig. 3 for the average D^0, D^+, D^{*+} D^0 meson, although in this case a higher number of binary nucleon-nucleon collisions is also expected to contribute to high-multiplicity events.

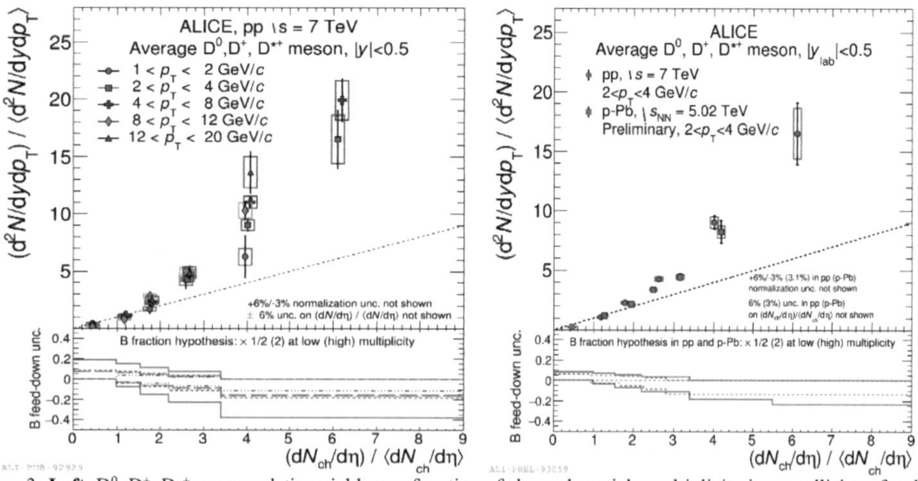

Figure 3: **Left**: D^0, D^+, D^{*+} meson relative yields as a function of charged-particle multiplicity in pp collisions for different p_T ranges at central rapidity. **Right**: self-normalized yield as a function of multiplicity in pp and p–Pb collisions for the average D^0, D^+, D^{*+} meson with $2 < p_T < 4$ GeV/c.

In Pb–Pb collisions, the open heavy-flavour R_{AA} measured with ALICE for D^0, D^+, D^{*+} [22-23] shows a strong reduction of the yields at large trasverse momenta ($p_T > 5$ GeV/c) in the most central collisions relative to a binary-scaled pp reference. This suppression is interpreted as due to charm quark in-medium energy loss. The expected mass ordering of the energy loss has been also investigated: Fig. 4 (left) shows the D-meson R_{AA} as a function of the average number of nucleons participating in the interaction, compared to the one of J/ψ from beauty-hadron decays measured by CMS [24]. The D-meson p_T range was chosen in order to obtain a significant overlap with the p_T distribution of B mesons decaying to J/ψ with $6.5 < p_T < 30$ GeV/c, thus allowing a consistent comparison. A similar trend as a function of centrality is observed, but the D-meson R_{AA} is systematically lower than the one of J/ψ from B decays. This is consistent with the expectation of a smaller in-medium energy loss for beauty than for charm quarks. Fig. 4 (right) shows also the comparison of the D-meson R_{AA} with that of charged hadrons and pions: a similar suppression is observed, but the uncertainties do not allow yet to draw a conclusion on the colour-charge and parton mass dependence of the in-medium energy loss.

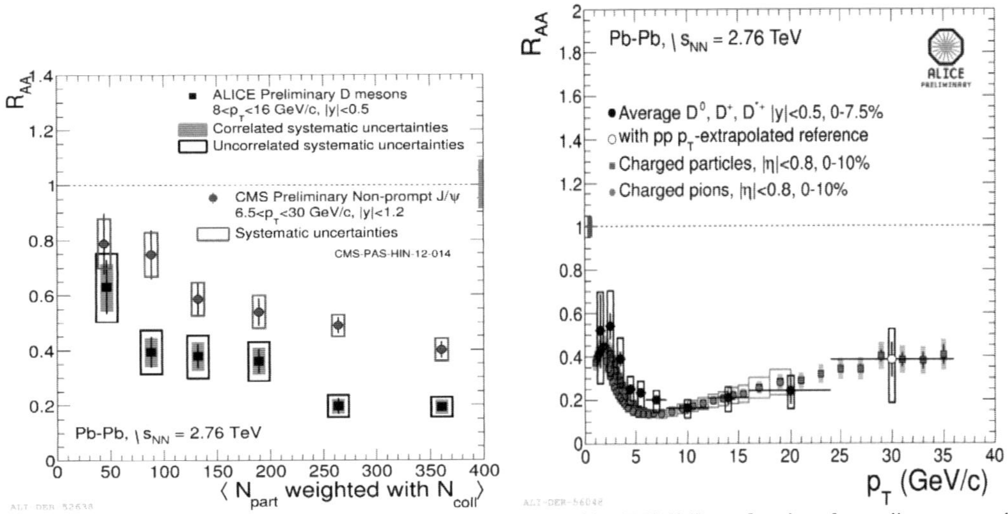

Figure 4: **Left**: R_{AA} of prompt D mesons and of non-prompt J/ψ measured by CMS [24] as a function of centrality, expressed in terms of the number of nucleons participating in the interaction. **Right**: Comparison of prompt D-meson R_{AA} with the R_{AA} of charged hadrons and pions.

The D-meson v_2 measured in 30-50% central Pb-Pb collisions shown in Fig. 5, is larger than zero with a 5.7σ significance in the interval $2 < p_T < 6$ GeV/c and comparable in magnitude to the one of charged hadrons, dominated by light-flavour hadrons [25] . These results indicate that at low p_T charm quarks participate in the collective motion of the system. At high p_T, v_2 results could give insight into the path-length dependence of the in-medium energy loss, but the present statistics does not allow to give a conclusion on this.

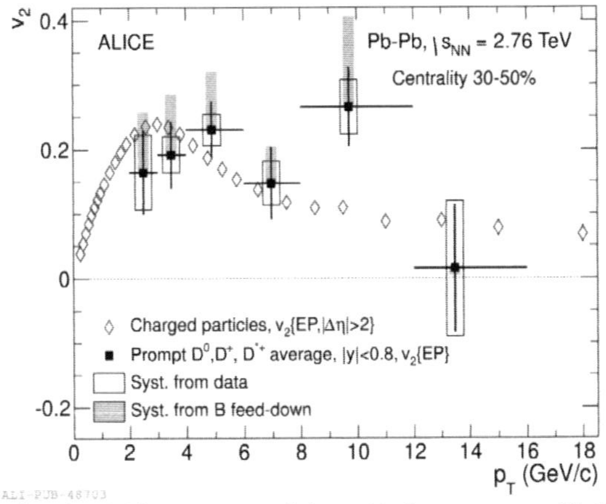

Figure 5: Comparison between D^0, D^{*+} and D^+ average v_2 and charged hadron v_2 measured in 30-50% central Pb-Pb collisions [25].

4. Conclusions

Measurements of open charm production have been carried out successfully with the ALICE experiment at the LHC. p_T-differential cross sections for prompt D mesons in pp collisions are reproduced within uncertainties by theoretical predictions based on pQCD calculations. The D-meson self-normalized yields exhibit an increasing trend with increasing charged-particle multiplicity in pp collisions, suggesting that MPIs affect the hard momentum scale relevant for heavy-flavour production. A similar increase is observed in p–Pb collisions, but in this case a higher number of binary nucleon-nucleon collisions also contributes to high-multiplicity events. R_{pPb} of prompt D mesons in p-Pb collisions is consistent with unity within uncertainties, providing evidence that CNM effects are small in the measured p_T range. In Pb-Pb collisions a strong suppression of heavy-flavour yields is observed at intermediate and high p_T. As the influence of CNM effects is small in the measurement p_T range, these results in Pb–Pb collisions can be interpreted as a final-state effect due to in-medium parton energy loss. The v_2 measured in Pb–Pb semi-central collisions is larger than zero at low p_T, suggesting that heavy quarks participate in the collective motion of the system.

References

[1] M. Gyulassy and M. Plumer, Phy. Lett. B243, 432 (1990).
[2] R. Baier, Y. L. Dokshitzer, A. H. Mueller, S. Peigne and D. Schiff, Nucl. Phys. B484, 265 (1997).
[3] M. H. Thoma and M. Gyulassy, Nucl. Phys. B351, 491 (1991).
[4] K. Eskola, H. Paukkunen and C. Salgado, JHEP 04 (2009) 065.
[5] D. de Florian and R. Sassot, Phys.Rev. D 69 (2004) 074028.
[6] I.Vitev et al., PRC 75 (2007) 064906.
[7] K. Aamodt et al., ALICE Collaboration, JINST 3 S08002 (2008).
[8] B. Abelev et al., ALICE Collaboration, JHEP 1201 (2012) 128.
[9] B. Abelev et al., ALICE Collaboration, Phys.Lett. B718 (2012) 279-294.
[10] B. Abelev et al., ALICE Collaboration, JHEP 1207 (2012) 191.
[11] M. Cacciari et al., CERN-PHTH/2011 227.
[12] G. Aad et al., ATLAS Collaboration, Eur.Phys. J. C 72 (2012).
[13] A. Adare et al., PHENIX Coll., Phys. Rev. Lett. 97 (2006) 252002.
[14] S. Alekhin et al., arXiv:hep-ph/0601013 (2006), chapter IV.
[15] B. Abelev et al., ALICE Collaboration, Phys.Rev.Lett. 113 (2014) 23, 232301.
[16] M.L. Mangano, P. Nason, G. Ridolfi, Nucl. Phys. B 373 (1992) 295.
[17] B. Abelev et al., ALICE Collaboration, JHEP 09 (2012) 112.
[18] R. Sharma, I. Vitev and B. Zhang, Phys. Rev. C 80 (2009) 054902.
[19] V. Khachatryan et al., CMS Collaboration, Eur. Phys. J. C 73 (2013) 2674.
[20] A. Abelev et al., ALICE Collaboration, JHEP 09 (2013) 049.
[21] J. Adam et al., ALICE Collaboration, arXiv:1505.00664 [nucl-ex] (2015).
[22] B. Abelev et al., ALICE Collaboration, PRL 109 112301 (2012).
[23] B. Abelev et al., ALICE Collaboration, JHEP 9 112 (2012).
[24] CMS Collaboration, CMS-PAS-HIN-12-014 (2013).
[25] B. Abelev et al., ALICE Collaboration, arXiv:1305.2707, [nucl-ex] (2013).

Journal of Physics: Conference Series **636** (2015) 012004 doi:10.1088/1742-6596/636/1/012004

J/ψ and $\psi(2S)$ measurements in p+p collisions at $\sqrt{s} = 500$ GeV in the STAR experiment

Qian Yang for the STAR Collaboration

University of Science and Technology of China, Heifei City, Anhui Province, China
Brookhaven National Laboratory, Upton, NY, USA

E-mail: ayqlyl@mail.ustc.edu.cn

Abstract. In this paper, we present the first measurement of J/ψ production in p+p collisions at $\sqrt{s} = 500$ GeV by the STAR experiment at RHIC in the rapidity range of $-1 < \mathrm{y} < 1$ and transverse momentum range of $4 < p_T < 20$ GeV/c. We also report the ratio of the integrated yield of $\psi(2S)$ to that of J/ψ over the p_T range of $4 < p_T < 12$ GeV/c.

1. Introduction

J/ψ production in heavy-ion collisions is an important tool for studying the properties of Quark-Gluon Plasma (QGP) [1]. Suppression of J/ψ production in heavy-ion collisions, due to color screening of quark and anti-quark potential, has been proposed as a signature of QGP formation [2]. Various measurements of J/ψ have been performed in different collision systems and at different energies, and indeed a suppression of J/ψ production has been observed [3,4,5,6]. Interpretation of the observed suppression requires a good understanding of the production mechanisms in elementary p+p collisions, which are however still largely unknown. Decay contributions from excited $c\bar{c}$ states and from B mesons could also complicate the picture [7]. J/ψ measurement at a new beam energy provides new insights on J/ψ production mechanisms and constrains theoretical model calculations. Furthermore, It is essential to push the measurement to high transverse momentum as model predictions differ the most at high p_T and many calculations are only applicable at high p_T.

2. Experiment and Analysis

The STAR detector is a large-acceptance multi-purpose detector which covers full azimuth and pseudo-rapidity of $|\eta| < 1$ [8]. A high tower trigger, utilizing Barrel Electromagnetic Calorimeter (BEMC) as the trigger detector, was used to select p+p collisions that potentially contained J/ψ. The data sample used for this analysis corresponds to an integrated luminosity of 22 pb^{-1}. The $J/\psi \rightarrow e^+e^-$ channel with a branching ratio of $B_{e^+e^-} = 5.9\%$ was utilized to reconstruct the J/ψ signals. Electrons were identified using the Time Projection Chamber (TPC) through ionization energy loss (dE/dx) measurement. At high p_T, BEMC can further reject hadrons via the E/p cut, in which E is the energy deposited in the BEMC, and p is the momentum of the track matched to the BEMC signal.

We obtained the J/ψ signal by subtracting the combinatorial and residual backgrounds from the unlike-sign e^+e^- invariant mass spectrum. Figure 1 shows the invariant mass distribution for di-electron pairs within $|y| < 1$. The unlike-sign pairs are represented by black histogram. The combinatorial background was reconstructed by the like-sign method, in which electrons (positrons) were paired with every other electrons (positrons) in the same event, shown as the filled histogram in Figure 1. After combinatorial background subtraction, the J/ψ candidates with residual background are shown as solid red circles. The residual background consists of contributions from Drell-Yan, $c\bar{c}$ pairs and $b\bar{b}$ pair decays. A Crystall Ball function plus an exponential function was used to simultaneously fit the J/ψ signal and the residual background. J/ψ yield was obtained by counting the bin contents after subtracting the residual background as well as the integral of the fitted Crystal Ball function in the mass range of $2.7 < M_{inv}^{ee} < 3.3$ GeV/c^2. The yields were corrected for the $\sim 10\%$ effect of $J/\psi \rightarrow e^+e^-\gamma$ decay that causes some of the J/ψ to be reconstructed outside the above mass range. Similar analysis procedures and criteria were used to extract $\psi(2S)$ signals. Figure 1 also shows the invariant mass distribution for $\psi(2S)$ in the insert. The $\psi(2S)$ raw yields were obtained in the mass range of $3.5 < M_{inv}^{ee} < 3.8$ GeV/c^2. Acceptance and efficiency corrections were studied using Monte Carlo (MC) embedding

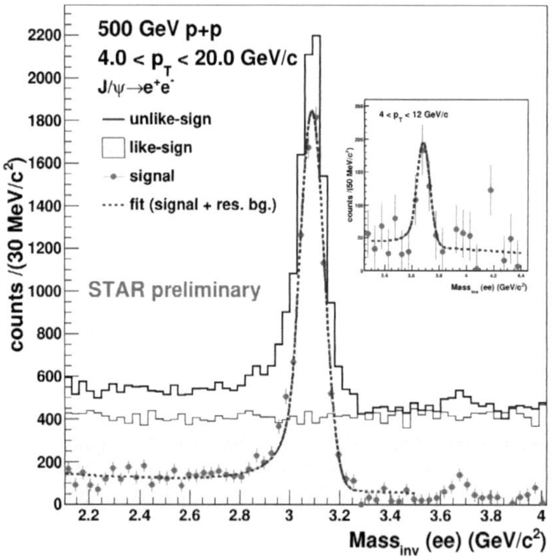

Figure 1. (Color online) Invariant mass distribution of unlike-sign (black histogram) and like-sign (filled histogram) electron/positron pairs, for $4 < p_T < 20$ GeV/c and $|y| < 1$. An exponential function plus a Crystall-ball function are used to fit the J/ψ signal and the residual background. The error bars depict the statistical error. $\psi(2S)$ signal is shown in the insert as well.

technique [11], in which simulated $J/\psi \rightarrow e^+e^-$ were mixed into real events at the raw data level. The systematic uncertainty in the procedure is estimated by varying kinematic and electron identification (eID) cuts in both data analysis and MC simulations coherently. We have also varied the invariant mass window and fitting method for signal counting to evaluate the systematic uncertainty in the yield extraction procedure, including the contributions from electron bremsstrahlung energy loss and correlated background. The systematic uncertainties are summarized in Tab. 1.

Table 1. Summary of systematic uncertainties of J/ψ and $\psi(2S)$ for different sources.

Description	J/ψ	$\psi(2S)/J/\psi$
Kinematic and eID cuts	6.1%	7.6%
Trigger efficiency	1.0%	3.0%
Yield extraction	1.3%	15%
Total	6.3%	17%

3. Results

Figure 2 shows the invariant cross section of J/ψ multiplied by the branching ratio at mid-rapidity ($|y| < 1$) as a function of p_T for p+p collisions at $\sqrt{s} = 500$ GeV. Other experimental results are also shown [12,13,14]. It has already been found that high p_T J/ψ follows the x_T scaling [12]: $E\frac{d^3\sigma}{dp^3} = g(x_T)/s^{\frac{n}{2}}$, where $x_T = 2p_T/\sqrt{s}$ and $n = 5.6 \pm 0.2$, which represents the number of constituents taking an active role in the production. We tested the x_T hypothesis with our measurement in Fig. 3, and we found that J/ψ follows x_T scaling at $p_T > 4$ GeV/c. While low p_T J/ψ production also originates from hard processes due to the mass scale, subsequent soft processes could cause a violation of x_T scaling.

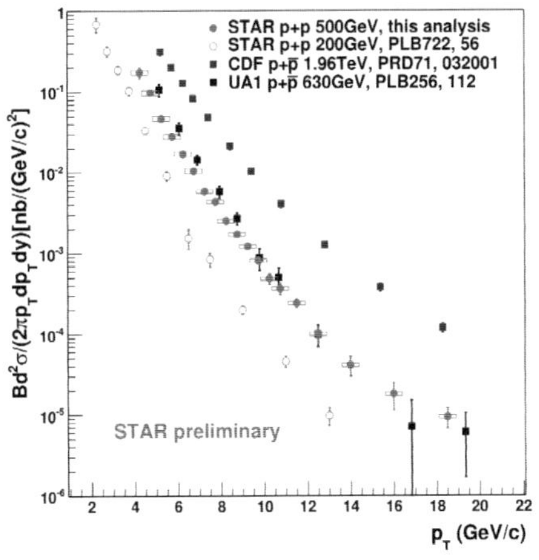

Figure 2. J/ψ p_T distributions measured at different energies. The error bars represent the statistical errors. The boxes represent the systematic uncertainties.

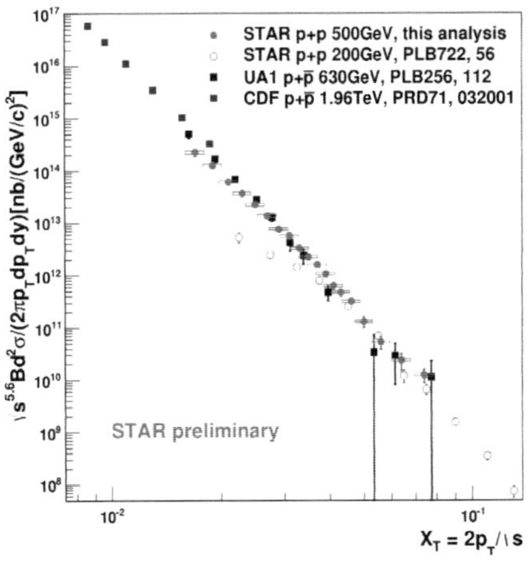

Figure 3. x_T distributions for J/ψ measured at different energies. The error bars represent the statistical errors. The boxes represent the systematic uncertainties.

The inclusive J/ψ production includes both prompt and non-prompt J/ψ. The prompt J/ψ production consists of the direct production ($\sim 60\%$) and feed-down contribution from excited states $\psi(2S)$ ($\sim 10\%$) and $\chi_C(\sim 30\%)$, while non-prompt J/ψ originate from B-meson decays. STAR has estimated the contribution from B-meson decays, within $10 \sim 25\%$ in the range of $4 < p_T < 12$ GeV/c, using azimuthal correlations between high p_T J/ψ and charged hadrons

[4]. We have measured the $\psi(2S)$ to J/ψ ratio in p+p collisions and the results are compared to other measurements at different colliding energies [15,16,17] in p+p or p+A collisions in Fig. 4. The STAR data point is consistent with the observed trend, with no collision energy dependence of the $\psi(2S)$ to J/ψ ratio seen within the current precision.

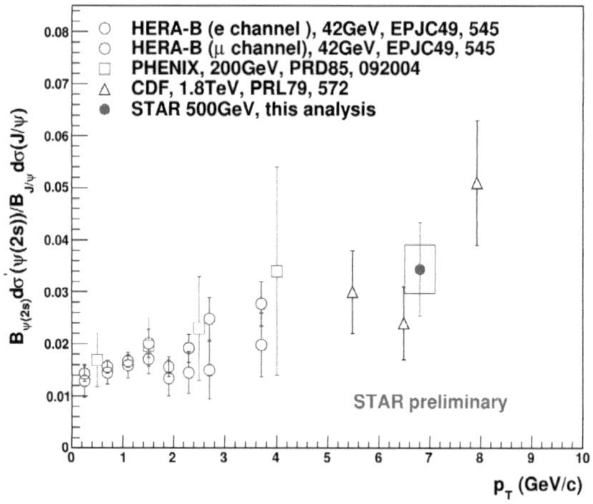

Figure 4. $\psi(2S)$ to J/ψ ratio, for $4 < p_T < 12$ GeV/c and $|y| < 1$. The error bars represent the statistical errors. The boxes represent the systematic uncertainties.

4. Summary

In summary, we report recent STAR measurements of the inclusive J/ψ cross section in p+p collisions at $\sqrt{s} = 500$ GeV, which extends p_T to 20 GeV/c. The first measurement of $\psi(2S)$ to J/ψ ratio is reported and compared with results from other experiments, with no collision energy dependence observed.

5. References

[1] M. Harrison, T.Ludlam and S.Ozaki, Nuclear Instruments and Methods inPhysics Research A 499 (2003) 235-244
[2] T. Matsui, H. Satz, Phys. Lett. B178 (1986) 416
[3] J. P. Lansberg, A Rev. Int. J. Mod. Phys., A21:3857, 2006
[4] M. C. Abreu et al, Phys. Lett. B, 449 :128, 1999
[5] R. Arnaldi et al, Phys. Rev. Lett., 99:132302, 2007
[6] E. T. Atomssa. 2008, arXiv:0805,4562v1
[7] N. Brambilla, et al., Eur. Phys. J. C 71 (2011) 1534
[8] Ackermannn K H et al, Nucl. Instr. Meth. A 499 624, 2003
[9] Beddo M et al, Nucl. Instr. Meth. A 499 725, 2003
[10] Anderson M et al, Nucl. Instr. Meth. A 499 659, 2003
[11] B.I. Abelev, et al., Phys. Rev. C 79 (2009) 034909
[12] L. Adamczyk et al, Phys. Lett. B, 722(2013)55-56
[13] D. Acosta et al, Phys. ReV. D 71, 032001 (2005)
[14] C. Albajar et al, Phys. Lett. B256(1991)112
[15] I. Abt et al, Eur. Phys. J. C49(2007)545-558
[16] A. Adare et al, Phys. Rev. D 85, 092004(2012)
[17] F. Abe et al, Phys. Rev. Lett. 79 572(1997)

Bottomia suppression in PbPb collisions at the LHC

Georg Wolschin
Institut für Theoretische Physik der Universität Heidelberg, Philosophenweg 16, D-69120 Heidelberg, Germany, EU
E-mail: g.wolschin@thphys.uni-heidelberg.de

Abstract. The suppression of Υ-mesons in the hot quark-gluon medium in PbPb collisions relative to scaled pp at $\sqrt{s_{NN}} = 2.76$ TeV is accounted for in a model that encompasses gluodissociation, collisional damping, screening, and reduced feed-down. Theoretical results for centrality-dependent suppression factors of the $\Upsilon(1S)$- and $\Upsilon(2S)$-states are compared with CMS and ALICE data. The measured triplet ground state suppression is well represented, but discrepancies between data and model persist for peripheral $\Upsilon(2S)$-events that call for further refinements, or additional suppression mechanisms.

1. Introduction

Whereas J/ψ-suppression in relativistic heavy-ion collisions has been investigated in detail for almost 30 years now, the research on the suppression of bottomia in the hot quark-gluon medium that is created in such collisions is a relatively new field. On the experimental side, it started with CMS data from the Large Hadron Collider (LHC) in PbPb collisions at 2.76 TeV c.m. energy per particle pair where the $\Upsilon(1S), \Upsilon(2S)$ and $\Upsilon(3S)$ states could be resolved for the first time in a heavy-ion collision ([1] and references therein), followed by the ALICE collaboration [2] – where the $\Upsilon(2S)$ and $\Upsilon(3S)$ states still wait for their resolution –, and STAR measurements from 200 GeV AuAu [3] at RHIC.

There are characteristic differences between J/ψ- and Υ-suppression which call for a detailed comparative investigation that is beyond the scope of this article. Let me mention, however, the large difference by more than a factor of three in the masses of charm and bottom quarks, and a corresponding difference in the meson masses ($m_{J/\psi} = 3.097$ GeV$/c^2$ vs. $m_{\Upsilon(1S)} = 9.460$ GeV$/c^2$). Accordingly in a PbPb collision at the present energy of 2.76 TeV more than 100 J/ψ-mesons are produced in initial hard collisions, but only about $6 - 10$ Υ-mesons and hence, statistical recombination of heavy quarks following the dissociation of the heavy mesons in the hot quark-gluon plasma (QGP) is an important issue in J/ψ-physics, but not in the Upsilon case. Although the statistical recombination of c- and $\bar{c}$-quarks is a most interesting process that may even be used as a possible QGP signature together with J/ψ-dissociation, its irrelevance in the $\bar{b}b$-case has the advantage of generating a cleaner environment for heavy-quarkonia dissociation in the QGP.

Another characteristic difference between J/ψ- and Υ-suppression is the unusual stability of the $\Upsilon(1S)$ ground state, which survives in the QGP up to very high temperatures. In particular, $\Upsilon(1S)$ resists dissociation due to screening of the real part of the quark-antiquark potential up to temperatures of more than four times the critical value, and similarly for other dissociation processes that we shall discuss in this note. This provides a very clean environment for investigating the centrality-dependent suppression of this state in the QGP which is quite different from both the excited Upsilon states, and all J/ψ-states.

The heavier the hadron that is produced in the collision, the shorter its formation time. Very heavy mesons such as the J/ψ, or the Υ meson in its $1S$ spin-triplet ground state are produced in hard collisions at very short times, typically at $\tau_F \simeq 0.1 - 0.5$ fm/c. Since the $\Upsilon(1S)$ state is particularly stable, it has a sizeable probability to survive in the hot quark-gluon medium that is produced in the fireball of a heavy-ion collision at LHC energies. We have accordingly devised a model that accounts for the dissociation of the various bottomium states in the hot medium (gluodissociation), the damping of the quark-antiquark binding due to the presence of the medium, which generates an imaginary part of the temperature-dependent potential, and the screening of the real part of the potential. The latter turns out to be unimportant for the $\Upsilon(1S)$ ground state, but it is relevant for the less strongly bound excited states.

Once the bottomia states have survived the quark-gluon plasma environment, the feed-down cascade from the excited states to the ground state has to be considered in detail. Due to the rapid depopulation of the excited states caused by the various mechanisms, the feed-down to the ground state is reduced accordingly, producing additional suppression relative to the situation in pp collisions at the same energy. We then compare the calculated centrality-dependent $\Upsilon(nS)$-suppression [4, 5] with recent CMS data [1], and predict the suppression at the LHC design energy.

2. Υ suppression in PbPb collisions at LHC energies

The production of heavy mesons and in particular, of bottomia in initial hard partonic interactions in relativistic PbPb collisions at LHC energies is of special interest, because quarkonia in the fireball can act as a probe to test the properties of the hot medium. In our model [4, 5] we investigate the suppression of the $\Upsilon(nS)$ states in PbPb as compared to the expectation from scaled pp collisions at the same energy, and compare with centrality-dependent CMS data [1] for the $1S$ and $2S$ states. In the hot medium, the three most important dissociation mechanisms are taken to be gluodissociation, damping, and screening of the real part of the quark-antiquark potential. The calculated gluodissociation cross section for the $1S$ and $2S$ states together with the thermal gluon distribution function are shown as functions of the gluon energy at two different temperatures in fig. 1.

Since the matrix elements that characterize gluodissociation (see formulae in [5]) have a finite extent in the gluon energy space, the corresponding cross sections exhibit maxima. Still, when averaged over the temperature-dependent gluon-energy distribution, and considering the momentum-dependent running of the strong-coupling constant [5], the average gluodissociation cross section rises monotonically as function of temperature in the region of interest – that is, for initial temperatures of $T \simeq 550$ MeV at the Υ formation time $\tau_F \simeq 0.1 - 0.5$ fm/c.

In addition to gluodissociation in the hot medium, we have also considered the damping widths due to the imaginary parts of the quark-antiquark potential, and the screening of the real parts. Gluodissociation and damping widths add up incoherently to total bottomia widths for the six states considered in the hot fireball, namely, $\Upsilon(1S, 2S, 3S)$ and $\chi_b(1P, 2P, 3P)$; other spin-triplet states are above the $B\bar{B}$ threshold and hence, are not relevant for a comparison with the measured $\Upsilon(nS)$ suppression.

The time and centrality dependence of the suppression of these six states in the hot medium is then calculated based on an ideal hydrodynamic time evolution of the medium that includes transverse expansion, as well as the relativistic Doppler effect whenever the mean transverse momenta of the produced bottomia states and the p_T of the expanding medium are different from each other. The temperature profiles obtained from our hydrodynamic calculation are displayed in fig. 2 for four different values of time in a central PbPb collision along the x^1-axis [5]. At the end of the QGP phase – when the temperature has dropped below the critical

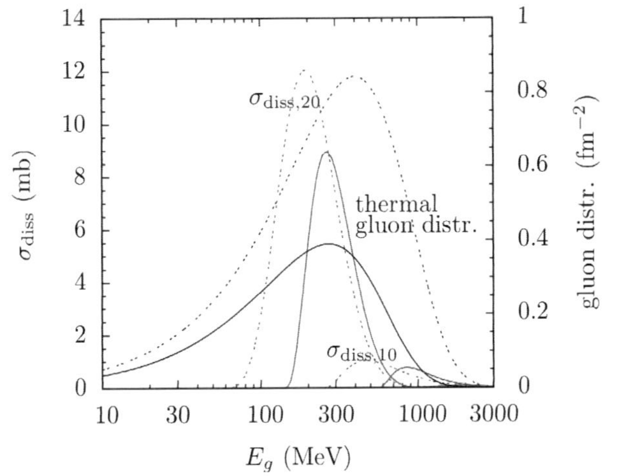

Figure 1. Gluodissociation cross section σ_{diss} (left scale) of the $\Upsilon(1S)$ and $\Upsilon(2S)$ states and the thermal gluon distribution (right scale) plotted for temperature $T = 170$ (solid curves) and 250 MeV (dotted curves) as functions of the gluon energy E_g. From Nendzig and Wolschin [5].

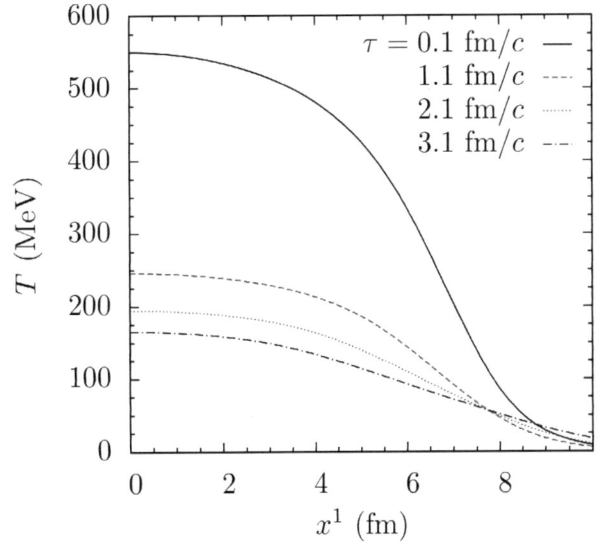

Figure 2. Calculated temperature profiles in the hydrodynamic expansion. From Nendzig and Wolschin [5].

temperature – we then calculate the centrality-dependent suppression factors in the QGP-phase of the collision, R_{AA}^{QGP}.

It is very relevant to include at all impact parameters the corona zone where the temperature never rises above the critical value so that none of the QGP dissociation mechanisms play a role there, leading to an occupation of the excited states even in more central collisions where the excited states melt in the high-temperature region due to screening. In the corona, additional hadronic suppression that is mostly due to collisions with pions – which we have investigated separately – may occur.

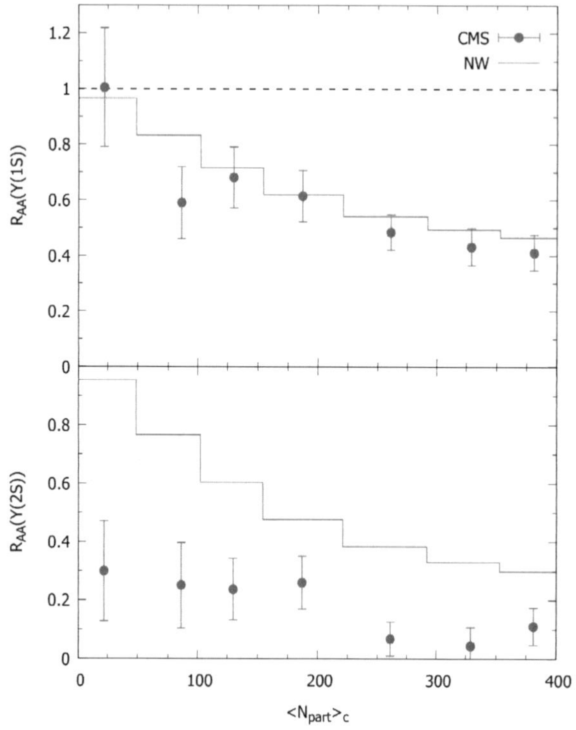

Figure 3. $\Upsilon(nS)$ suppression factors in 2.76 TeV PbPb measured by CMS [1] (circles) and calculated with the model of Nendzig and Wolschin [5] (solid lines) for $\tau_{init} = 0.1$ fm/c, $\tau_F = 0.5$ fm/c, $T_i = 550$ MeV and $T_c = 170$ MeV, averaged over transverse momentum, 4 GeV/$c < p_T <$ 24 GeV/c. From [6].

Once the centrality-dependent suppression in the hot fireball and in the corona is established, the bottomia states de-excite in a feed-down cascade before being detected through the emission of $\mu^+\mu^-$ pairs, with a branching ratio of approximately 2.48 % for the $1S$ state, 1.93 % for the $2S$ state, and 2.18 % for the $3S$ state. The calculation of the feed-down cascade including the newly detected $\chi_b(3P)$ state is described in [8]. It yields the final centrality-dependent suppression factors R_{AA} [5] as shown in fig. 3 in comparison with the CMS data [1] for the $\Upsilon(1S)$ and $\Upsilon(2S)$ states for a parameter set indicated in the caption.

The suppression of the spin-triplet ground state and its centrality dependence is in good agreement with the data for all centralities except for the $40-50\%$ bin. For the first excited $2S$ state, however, the calculated centrality dependence is too steep. Although the overall amount of suppression would be enhanced through a shorter formation time τ_F with ensuing larger QGP lifetime, such that the central bins are better accounted for, the problem of missing suppression in the three peripheral bins $50-100$ %, $40-50$ % and $30-40$ % remains to be solved. We have checked the influence of hadron-induced dissociation in the corona region, with the result that the centrality-dependence of pion-induced dissociation is probably not strong enough to account for the difference. Another interesting possibility may be the dissociation through the transient magnetic fields, which is expected to be more pronounced for peripheral collisions.

The suppression depends quite strongly on the transverse momentum of the produced bottomia. In our early work [4] we had assumed bottomia production at rest, whereas in [5]

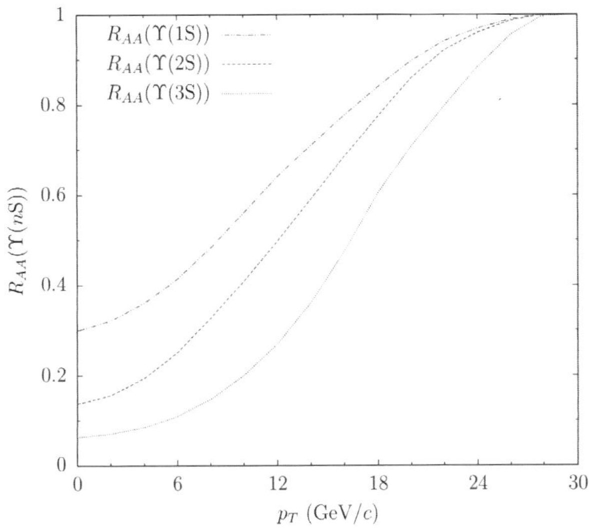

Figure 4. Transverse-momentum dependence of the suppression factors $R_{AA}(1S)$, $R_{AA}(2S)$ and $R_{AA}(3S)$ in central PbPb collisions at $\sqrt{s_{NN}} = 2.76$ TeV plotted for an Υ formation time of $\tau_F = 0.3$ fm/c. From [7].

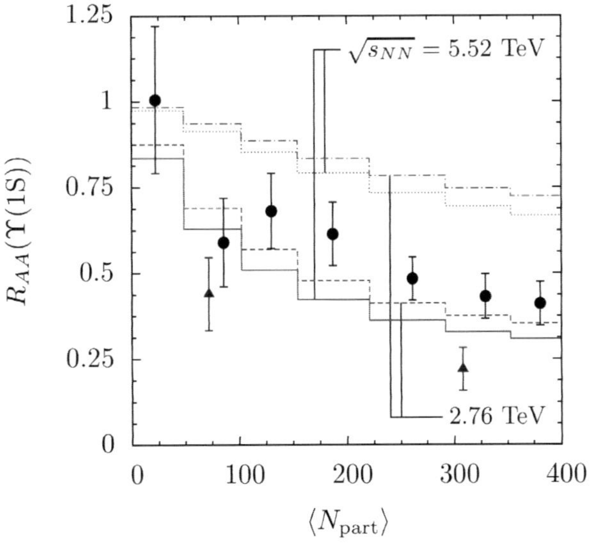

Figure 5. Predictions for $R_{AA}(1S)$ and $R_{AA}^{QGP}(1S)$ for PbPb collisions at $\sqrt{s_{NN}} = 5.52$ TeV (solid and dotted lines, respectively) and the previous results for $\sqrt{s_{NN}} = 2.76$ TeV (dashed and dash-dotted lines, respectively) as a function of centrality, averaged over p_T, plotted for an Υ formation time of $\tau_F = 0.3$ fm/c, together with data from CMS [1] and ALICE [2] for $\sqrt{s_{NN}} = 2.76$ TeV. From [5].

we have investigated the more realistic case of finite transverse momenta of the bottomia. The collisions with the light medium particles do not substantially change the transverse momenta of the bottomia and hence, there will be a finite relative velocity between them and the expanding QGP. The result is a relativistic Doppler shift which causes an angle-dependent effective temperature with a blue shift in the forward, and a red shift in the backward direction relative to the movement of the heavy meson in the medium. With increasing velocity, the red-shifted region is growing, whereas the blue-shifted region is confined to increasingly smaller angles. Averaging over the solid angle, we find that the red shift has more impact on the result, reducing thus the effective temperature that is decisive for the action of the dissociation processes.

The corresponding p_T-dependence of the suppression factor for the three $\Upsilon(nS)$-states is displayed in fig. 4. The suppression is most pronounced for $p_T < 6$ GeV/c and is diminished for large values of p_T, in qualitative agreement with the expectation from the data. There is, however, no mechanism in our model that could yield a suppression factor above one at large transverse momenta. In the calculation of the suppression shown in the previous factor, we have averaged over the p_T-dependence of the suppression factor in the range 4 GeV/$c \leq p_T \leq 24$ GeV/c. We have omitted lower p_T-values because the CMS data cover $p_T > 4$ GeV/c.

Since the results for the ground-state suppression are in agreement with experiment, we have also calculated the $\Upsilon(1S)$ suppression in PbPb at the LHC design energy of 5.52 TeV, solid line in fig. 5. The central initial temperature is increased by 6.6 % to 586 MeV using the scaling relation between the initial entropy density and the charged-particle multiplicity density $s_0 \propto dN_{ch}/d\eta \propto T_0^3$. The ground state is found to be slightly more suppressed in the medium at the higher energy and hence, the total suppression in PbPb as compared to *pp* is stronger. However, the additional ground state suppression effect is less than 10 % when doubling the c.m. energy.

3. Conclusion

The suppression of the strongly bound $\Upsilon(1S)$ spin-triplet ground state in PbPb collisions at LHC energies as compared to scaled *pp* collisions at the same energy is a sensitive indicator for the properties of the quark-gluon plasma that causes gluodissociation and damping. The feed-down cascade from the excited bottomia states to the ground state produces additional suppression, since the excited states melt through screening, or depopulate through dissociation processes and hence, there is less feed-down to the $1S$ ground state as compared to *pp* at the same energy.

Our model results [5] for the ground state are in good agreement with the CMS data [1]. The transverse-momentum dependence of the modification factor still needs to be compared with forthcoming data and probably requires further investigation, and also the suppression of the first excited state calls for additional centrality-dependent dissociation mechanisms.

I am grateful to my students – in particular, Felix Nendzig and Johannes Hoelck – for their contributions, and to René Bellwied and the organizers of the Winter Workshop on Nuclear Dynamics 2015 for setting up this interesting conference in Colorado, and for their hospitality. Figs. 1, 2 and 5 are © IOP Publishing, all rights reserved. This work has been partially supported by IMPRS-PTFS, the ExtreMe Matter Institute EMMI, and DFG through TRR-33.

References

[1] S. Chatrchyan et al. (CMS Collaboration), Phys. Rev. Lett. **109**, 222301 (2012)
[2] B. Abelev et al. (ALICE Collaboration), Phys. Lett. B **738**, 361 (2014)
[3] L. Adamczyk et al. (STAR Collaboration), Phys. Lett. B **735**, 127 (2014)
[4] F. Nendzig, G. Wolschin, Phys. Rev. C **87**, 024911 (2013)
[5] F. Nendzig, G. Wolschin, J. Phys. G: Nuclear and Particle Physics **41**, 095003 (2014)
[6] J. Hoelck, G. Wolschin, unpublished (2014)
[7] F. Nendzig, G. Wolschin, unpublished (2015)
[8] F. Vaccaro, F. Nendzig, G. Wolschin, Europhys. Lett. **102**, 42001 (2013)

31st Winter Workshop on Nuclear Dynamics (WWND2015) IOP Publishing
Journal of Physics: Conference Series **636** (2015) 012006 doi:10.1088/1742-6596/636/1/012006

A unified description of the reaction dynamics : Comparing pPb and PbPb collisions at the LHC

Klaus Werner
SUBATECH, University of Nantes – IN2P3/CNRS– EMN, Nantes, France

Abstract. There is little doubt that in heavy ion collisions at the LHC, we observe a hydrodynamically expanding system, providing strong evidence for the formation of a Quark Gluon Plasma (QGP) in the early stage of such collisions. These observations are to a large extent based on results on azimuthal anisotropies, perfectly compatible with a hydrodynamic evolution. Surprisingly, in p-Pb collisions one observes a very similar behavior. Motivated by the experimentally observed similarities of v_n results in pPb and PbPb collisions at similar multiplicities at the LHC [1], we analyze these findings using the EPOS3 approach.

1. Introduction

Collective hydrodynamic flow seems to be well established in heavy ion (HI) collisions at energies between 200 and 2760 AGeV, whereas p-p and p-A collisions are often considered to be simple reference systems, showing "normal" behavior, such that deviations of HI results with respect to p-p or p-A reveal "new physics". Surprisingly, recent results from p-Pb at 5 TeV on the transverse momentum dependence of azimuthal anisotropies are very similar to the observations in HI scattering at comparable multiplicities.

In fig. 1, we show results from the CMS collaboration[1] on the p_t dependence of v_2 of charged

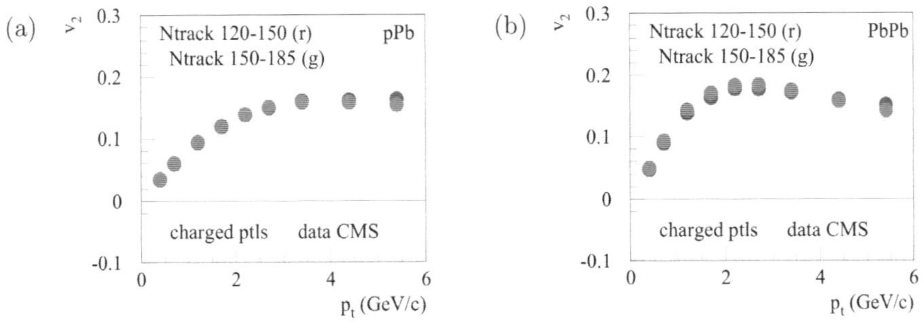

Figure 1. (Color online) p_t dependence of v_2 of charged particles from the CMS collaboration[1].

particles for two different multiplicity ranges, for pPb (a) and PbPb (b). The results for the two systems are quite similar, although the shapes in pPb and PbPb are somewhat different. We

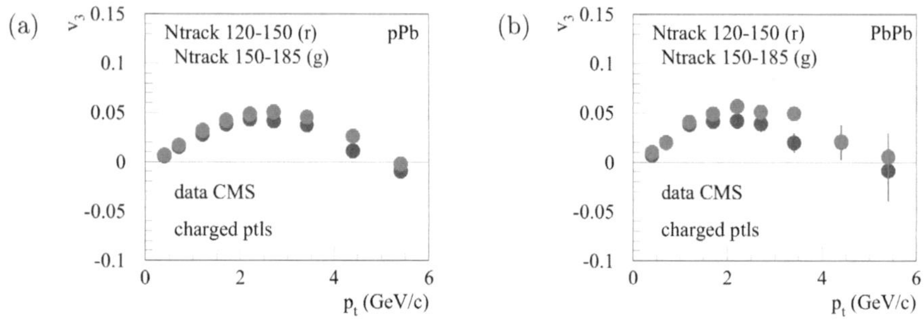

Figure 2. (Color online) p_t dependence of v_3 of charged particles from the CMS collaboration[1].

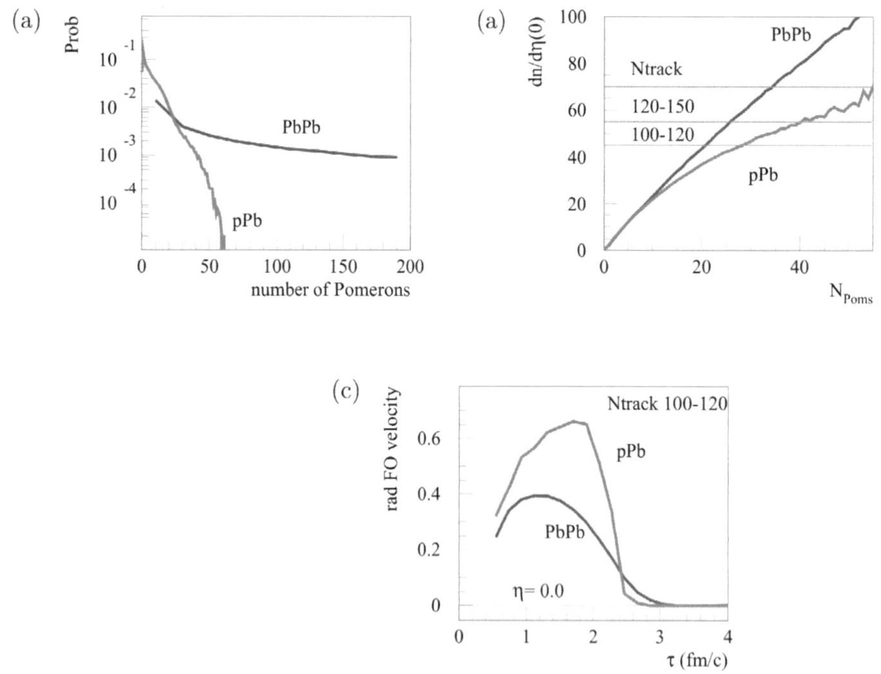

Figure 3. (Color online) (a) Pomeron number distribution in pPb and PbPb collisions (the PbPb distribution extends up to about 2000). (b) Average charged particle multiplicity density at central pseudorapidity $dn/d\eta(0)$ as a function of the Pomeron number. (c) Radial flow at freeze-out in pPb and PbPb for a given N_{track} range at space-time rapidity $\eta = 0$.

observe for both systems little change with multiplicity. Remarkable is the fact that v_2 at large p_t (where hard physics is expected to dominate) remains large.

In fig. 2, we show results from the CMS collaboration[1] on the p_t dependence of v_3 of charged particles for the two different multiplicity ranges, for pPb (a) and PbPb (b). Here, the

Journal of Physics: Conference Series **636** (2015) 012006 doi:10.1088/1742-6596/636/1/012006

two systems are almost perfectly identical. The curves increase slightly with Ntrack. For large values of p_t, in all cases, v_3 becomes small.

In [1] they study as well the mass splitting in the p_t dependence of v_2 and v_3 for kaons compared to lambdas. Again, pPb and PbPb show the same effect.

In EPOS 3 [2], the basic quantity to characterize the geometry of a collision is the number $N_{\rm Pom}$ of Pomerons. In fig. 3(a), we plot the $N_{\rm Pom}$ distributions in pPb and PbPb. The overlap area amounts to around 60 or less Pomerons. The Pomeron number is very closely related to multiplicity, but in a different way for pPb and PbPb, as shown in fig.3(b), where we plot the average charged particle multiplicity density at central pseudorapidity $dn/d\eta(0)$ as a function of the Pomeron number. We also indicate the intervals corresponding to the N_{track} ranges used by CMS. The corresponding Pomeron numbers are systematically higher for pPb compared to PbPb. The reason is that the radial flow at freeze-out is much bigger in pPb, as shown in fig.3(c), where we plot the radial flow at freeze-out in pPb and PbPb for a given N_{track} range. This is due to the fact that in pPb the initial radial size is very small leading to a fast radial expansion. This strong flow in pPb has visible consequences.

2. Unified description of the reaction dynamics

To analyze the p_t dependence of v_2 and v_3 in given given multiplicity ranges for pPb and PbPb collisions, we employ EPOS3 [2], a unified description of the dynamics of ALL reactions, from p-p to AA. EPOS3 [2] is based on several stages:

Initial conditions. A Gribov-Regge multiple scattering approach is employed [3], where the elementary object (by definition called Pomeron) is a DGLAP parton ladder, using in addition a CGC motivated saturation scale [5] for each Pomeron, of the form $Q_s \propto N_{\rm part}\,\hat{s}^{\lambda}$, where $N_{\rm part}$ is the number of nucleons connected the Pomeron in question, and $\hat{s}$ its energy. The parton ladders are treated as classical relativistic (kinky) strings. Our formalism is referred to as "Parton-Based Gribov-Regge Theory" (PBGRT).

Core-corona approach. At some early proper time τ_0, one separates fluid (core) and escaping hadrons, including jet hadrons (corona), based on the momenta and the density of string segments [4, 2]. The corresponding energy-momentum tensor of the core part is transformed into an equilibrium one, needed to start the hydrodynamical evolution. This is based on the hypothesis that equilibration happens rapidly and affects essentially the space components of the energy-momentum tensor.

Viscous hydrodynamic expansion. Starting from the initial proper time τ_0, the core part of the system evolves according to the equations of relativistic viscous hydrodynamics [2, 6], where we use presently $\eta/s = 0.08$. A cross-over equation-of-state is used, compatible with lattice QCD [7, 8].

Statistical hadronization The "core-matter" hadronizes on some hypersurface defined by a constant temperature T_H, where a so-called Cooper-Frye procedure is employed, using equilibrium hadron distributions, see [8].

Final state hadronic cascade After hadronization, the hadron density is still big enough to allow hadron-hadron rescatterings. For this purpose, we use the UrQMD model [9].

The above procedure is employed for each event (event-by-event procedure).

Whereas our approach is described in detail in [2], referring to older works [3, 4, 8], we confine ourselves here to a couple of remarks, to selected items. The initial conditions are generated in the Gribov-Regge multiple scattering framework. Our formalism is referred to as "Parton-Based Gribov-Regge Theory" (PBGRT) and described in very detail in [3], see also [2] for all the details of the present (EPOS3) implementation. The fundamental assumption of the approach is the hypothesis that the S-matrix is given as a product of elementary objects, referred to

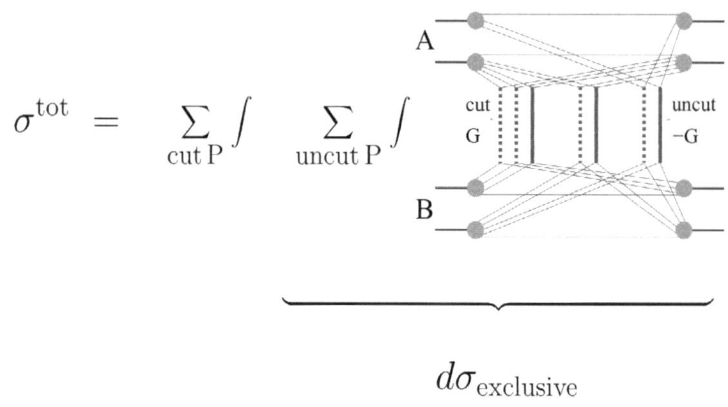

Figure 4. (Color online) PBGRT formalism: The total cross section expressed in terms of cut (dashed lines) and uncut (solid lines) Pomerons, for nucleus-nucleus, proton-nucleus, and proton-proton collisions. Partial summations allow to obtain exclusive cross sections, the mathematical formulas can be found in [3], or in a somewhat simplified form in [2].

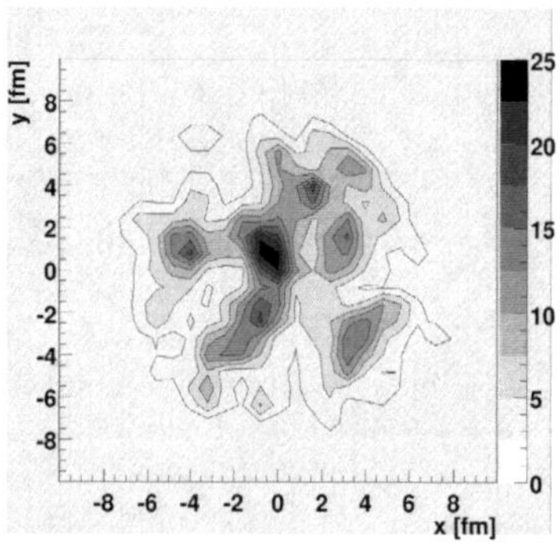

Figure 5. (Color online) Initial condition for QGP as obtained in the PBGRT framework in the year 2000, see ref [10].

as Pomerons. Once the Pomeron is specified (taken as a DGLAP parton ladder, including a saturation scale), everything is completely determined. Employing cutting rule techniques, one may express the total cross section in terms of cut and uncut Pomerons, as sketched in fig. 4. The great advantage of this approach: Doing partial summations, one obtains expressions for partial cross sections $d\sigma_{\text{exclusive}}$, for particular multiple scattering configurations, based on which the Monte Carlo generation of configurations can be done. No additional approximations are needed. The above multiple scattering picture is used for p-p, p-A, and A-A.

Based on the PBGRT approach, we obtain in A-A collisions a very large number of strings, but the randomness of their transverse positions leads to "bumpy" energy density distributions in the transverse plane at τ_0, as published for the first time in the year 2000, see fig. 21 of [10], reproduced as fig. 5 in this paper.

To understand the results later in this paper, we will discuss an example of core-corona separation in a semi-peripheral p-Pb collision, as shown in fig. 6. Shown (in the left figure)

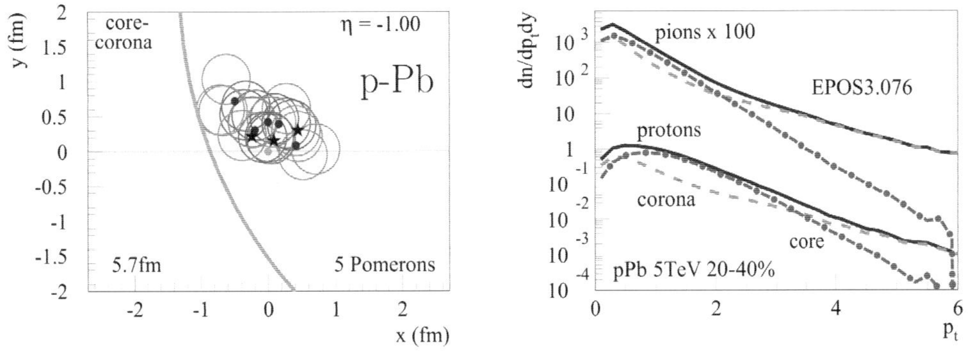

Figure 6. (Color online) Left: Core-corona separation in a semi-peripheral p-Pb collision. Right: the corresponding core and corona contributions to the p_t spectra of pions and protons.

are string segments in the transverse plane, red (core) and green (corona) ones. There are sufficient overlapping core string segments to provide a core of plasma matter, showing a (short) hydrodynamic expansion, quickly building up flow. In the right figure, we plot the contribution from core and corona to the p_t spectra of pions and protons. In particular for protons, the core dominates at intermediate p_t (mass effect).

3. Results for v_2

Coming back to the centrality dependence of v_n, we plot in fig. 7(a) the p_t dependence of v_2 for

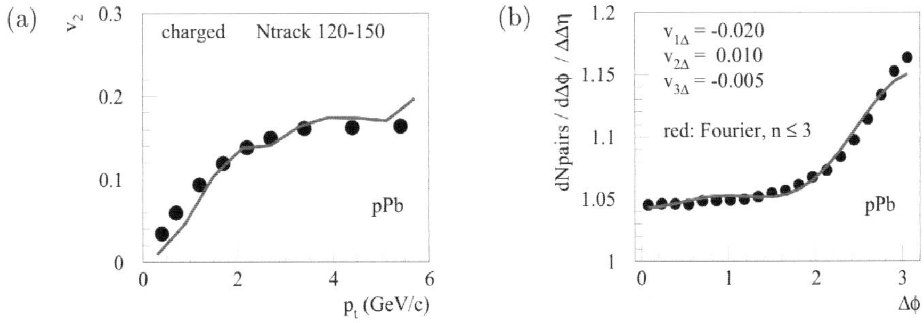

Figure 7. (Color online) (a) p_t dependence of v_2 for charged particles from EPOS3 without hydro (red curve)) compared to CMS data[1] (points). (b) Two hadron correlation function from EPOS3 without hydro (black points) together with the Fourier series based on three coefficients (plus the zeroth one).

charged particles from EPOS3 without hydro (red curve) compared to CMS data[1] (points). Amazingly the EPOS curve reproduces the data, although in the model there is no hydrodynamic flow active, the effect is a pure jet effect. To better see this, we check the dihadron correlation function, with p_t between 1 and 2 GeV/c, integrated over $\Delta\eta$ with $|\Delta\eta| > 2$, as shown in fig. 7(b). We plot the two hadron correlation function from EPOS3 without hydro (black points) together with the Fourier series based on three coefficients (plus the zeroth one), which already reproduces the correlation function quite well. Considering the coefficients, we note that the third one ($v_{3\Delta}$) is negative, in complete disagreement with experimental data! So one can say that the second coefficient agrees "by accident".

In the following, we employ full EPOS (including hydro and hadronic cascade). In fig. 8,

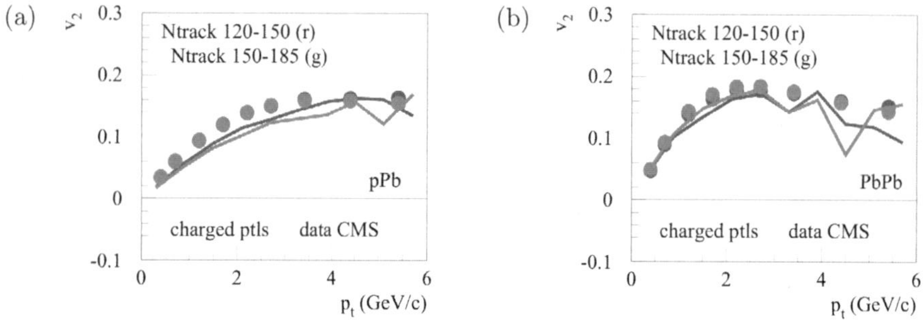

Figure 8. (Color online) p_t dependence of v_2 for charged particles from EPOS3 (lines) compared to CMS data[1] (points), for two different multiplicity ranges, in (a) pPb and (b) PbPb collisions.

we plot the p_t dependence of v_2 for charged particles from EPOS3 (lines) compared to CMS data[1] (points), for two different multiplicity ranges, in (a) pPb and (b) PbPb collisions. We observe for the calculation the same trend as seen in the data: Little change with multiplicity, large v_2 at large p_t, similar magnitude in pPb and PbPb, but different shape, understood in the model to be due to the bigger radial flow in pPb compared to PbPb. Why do we nevertheless obtain similar magnitudes? This is due to the fact that the smaller flow in PbPb is compensated by bigger eccentricities ϵ_2, as shown in fig. 9, where we plot the eccentricities ϵ_n from EPOS3,

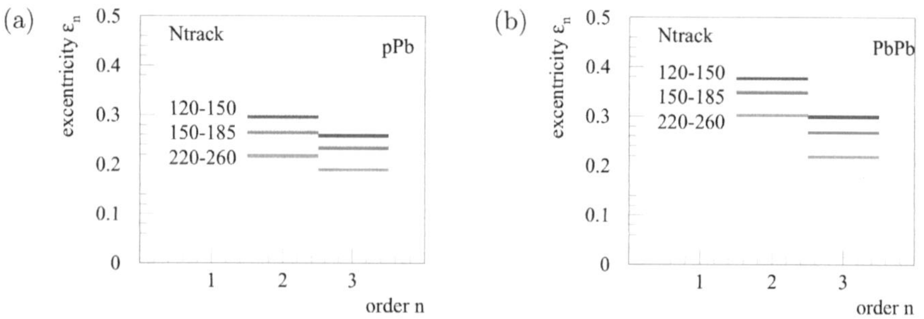

Figure 9. (Color online) Eccentricities ϵ_n from EPOS3, for three different multiplicity ranges, in (a) pPb and (b) PbPb collisions.

for three different multiplicity ranges, in (a) pPb and (b) PbPb collisions.

To summarize: Comparing pPb and PbPb, for the same multiplicities, using EPOS 3.111 (optimized to get pt spectra), we find: The p_t dependence of v_2 for the two systems is quite similar (as in the data). Microscopic properties are different though: There is more flow in pPb, compensated by bigger eccentricities in PbPb.

[1] CMS collaboration, Phys. Lett. B 742 (2015) 200, arXiv:1409.3392
[2] K. Werner et al., Phys.Rev. C89 (2014) 064903, arXiv:1312.1233
[3] H. J. Drescher, M. Hladik, S. Ostapchenko, T. Pierog and K. Werner, Phys. Rept. 350, 93, 2001
[4] K. Werner, Phys. Rev. Lett. 98, 152301 (2007)
[5] L. McLerran, R. Venugopalan, Phys. Rev. D 49 (1994) 2233; L. McLerran, R. Venugopalan, Phys. Rev. D 49 (1994) 3352; L. McLerran, R. Venugopalan, Phys. Rev. D 50 (1994) 2225.
[6] Iu. Karpenko, P. Huovinen, M. Bleicher, arXiv:1312.4160
[7] S. Borsanyi et al., JHEP 1011 (2010) 077, arXiv:1007.2580
[8] K. Werner, Iu. Karpenko, T. Pierog, M. Bleicher, K. Mikhailov, arXiv:1010.0400, Phys. Rev. C 83, 044915 (2011)
[9] M. Bleicher et al., J. Phys. G25 (1999) 1859; H. Petersen, J. Steinheimer, G. Burau, M. Bleicher and H. Stocker, Phys. Rev. C78 (2008) 044901
[10] H.J. Drescher, S. Ostapchenko, T. Pierog, K. Werner, hep-ph/0011219, PhysRevC.65.054902

Monte-Carlo approach to particle-field interactions and the kinetics of the chiral phase transition

Carsten Greiner[1], Christian Wesp[1], Hendrik van Hees[1,2], Alex Meistrenko[1]

[1]Institut für theoretische Physik, Goethe-Universität Frankfurt am Main, Max-von-Laue-Straße 1, D-60438 Frankfurt, Germany

[2]Frankfurt Institute for Advanced Studies, Ruth-Moufang-Straße, D-60438 Frankfurt, Germany

E-mail: Carsten.Greiner@th.physik.uni-frankfurt.de

Abstract. The kinetics of the chiral phase transition is studied within a linear quark-meson-σ model, using a Monte-Carlo approach to semiclassical particle-field dynamics. The meson fields are described on the mean-field level and quarks and antiquarks as ensembles of test particles. Collisions between quarks and antiquarks as well as the $q\overline{q}$ annihilation to σ mesons and the decay of σ mesons is treated, using the corresponding transition-matrix elements from the underlying quantum field theory, obeying strictly the rule of detailed balance and energy-momentum conservation. The approach allows to study fluctuations without making ad hoc assumptions concerning the statistical nature of the random process as necessary in Langevin-Fokker-Planck frameworks.

1. Introduction

One of the motivations for the study of ultrarelativistic heavy-ion collisions is to gain a detailed understanding of the phase diagram of strongly interacting matter [1]. At the largest energies as achieved at the Large Hadron Collider (LHC) and the Relativistic Heavy Ion Collider (RHIC) a hot and dense fireball is formed which can be described to a surprising accuracy as a nearly perfect fluid of strongly coupled quarks and gluons (QGP) undergoing a transition to a hot hadron-resonance gas. In these situations, where the net-baryon density or the baryon-chemical potential are small, lattice-QCD (lQCD) calculations indicate a crossover transition from confined to deconfined matter as well as from a phase where chiral symmetry is spontaneously broken to one where it is restored at a (common) transition temperature $T_{\rm c} \simeq 160$ MeV [2].

At lower collision energies, as studied in the RHIC beam-energy scan (BES) program, at the CERN SPS, and the future FAIR and NICA experiments, the produced medium starts at lower temperatures and larger net-baryon densities. Since in this situation the application of lQCD is challenging due to the "sign problem" at finite $\mu_{\rm B}$, one relies on effective chiral models, which predict the existence of a first-order chiral-phase-transition line ending in a critical point of a second-order phase transition.

For theory the challenge is to provide possible observables for this phase structure in heavy-ion collisions, like the ("grand-canonical") fluctuations of conserved charges like net-baryon number or electric charge. Not only the question, how to effectively model the phase transition (e.g., with the Nambu-Jona-Lasinio (NJL) or the (linear) σ model with extensions taking into account gluonic degrees of freedom implementing Polyakov loops) arises but also, which of the features of the phase structure predicted for such models applying thermal quantum field theory (describing a medium in thermal and chemical equilibrium) like (critical) fluctuations of conserved charges survive for a rapidly expanding and cooling fireball as created in heavy-ion collisions.

To address the latter question, one relies on transport simulations to describe the off-equilibrium dynamics of the fireball. One approach is the use of ideal or viscous hydrodynamics to describe the bulk evolution of the fireball (assuming a state close to local thermal equilibrium), which successfully describes key phenomena of heavy-ion collisions, and adding the fluctuations by hand in a Langevin approach [3–5]. On the other hand this implies that the statistics of the random process has to be put in as an ad hoc assumption. Usually a Gaussian Markovian ("white noise") is assumed, but the simulation of non-Markovian ("colored noise") processes is feasible in principle [6].

On the other hand, one would like to study the implication of different transition scenarios (cross-over, 1st order, 2nd order) as realized in the various quantum-field theoretical models on the nature of the hopefully observable fluctuations [7–13] on the statistics of the probably observable fluctuations.

In this work we present a novel Monte-Carlo approach to address this challenging problem using the most simple quark-meson linear σ model [14]. The meson fields are treated on the mean-field level and the quarks and antiquarks are realized in terms of a test-particle ensemble. Here the challenge is to implement "discrete" local interaction processes like elastic collisions and reactions like the $q\overline{q}$ annihilation to a σ meson and the decay of σ mesons to a $q\overline{q}$ pair admitting not only kinetic but also chemical equilibration starting from an off-equilibrium situation, using the transition-probability matrix elements of the underlying quantum field theory.

2. Linear quark-meson σ model

To investigate the feasibility of a kinetic description of the off-equilibrium dynamics of the chiral phase transition the most simple two-flavor chiral model, based on the chiral group $\mathrm{SU}(2)_{\mathrm{L}} \times \mathrm{SU}(2)_{\mathrm{R}}$ is considered, using a flavor doublet of Dirac fields ψ, describing u and d quarks and antiquarks and a four-dimensional real-valued set of scalar fields $(\sigma, \vec{\pi})$ transforming under the SO(4) representation of the chiral group [15]. The Lagrangian reads

$$\mathcal{L} = \overline{\psi}[\mathrm{i}\not{\partial} - g(\sigma + \mathrm{i}\gamma_5 \vec{p} \cdot \vec{\tau})]\psi + \frac{1}{2}(\partial_\mu \sigma \partial^\mu \sigma + \partial_\mu \vec{\pi} \cdot \partial^\mu \vec{\pi}) - U(\sigma, \vec{\pi}) \tag{1}$$

with the meson potential

$$U(\sigma, \vec{\pi}) = \frac{\lambda^2}{4}(\sigma^2 + \vec{\pi}^2 - \nu^2)^2 - f_\pi m_\pi^2 \sigma - U_0, \tag{2}$$

where $g \in [3.3, 5.5]$ denotes the Yukawa coupling between quarks and mesons, $\lambda^2 = 20$ the meson coupling constant (corresponding to a σ mass of $m_\sigma \simeq 600$ MeV), f_π the pion-decay constant, and $\nu^2 = f_\pi^2 - m_\pi^2/\lambda^2$. The potential (2) contains the explicit breaking of the chiral symmetry due to the finite current quark masses resulting in a non-zero pion mass, $m_\pi \simeq 138$ MeV, of the pseudo-Goldstone modes $\vec{\pi}$. The constituent quark masses are given by $m_q^2 = g^2 \sigma_0^2$.

The grand-canonical potential in mean-field approximation reads

$$\Omega(T, \mu) = U(\sigma, \vec{\pi}) + \Omega_{\overline{\psi}\psi} \tag{3}$$

with

$$\Omega_{\overline{\psi}\psi} = -d_n \int \frac{\mathrm{d}^3\vec{p}}{(2\pi)^3} \left[E + T\ln(1 + \exp(-\beta(E - \mu))) + T\ln(1 + \exp(-\beta(E + \mu)))\right], \tag{4}$$

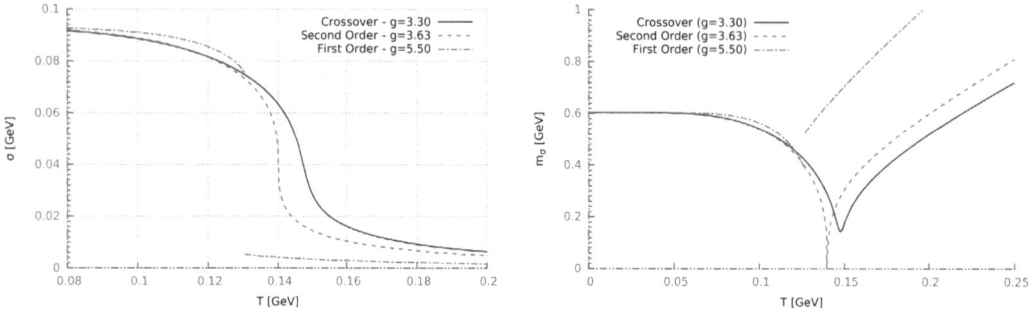

Figure 1. The phase diagram for the linear σ model in mean-field approximation at $\mu_{\mathrm{B}} = 0$, $\langle\vec{\pi}\rangle = 0$. Left: the order parameter $\langle\sigma\rangle$ and the effective σ mass $m_\sigma = \partial^2\Omega/\partial\sigma^2$.

where $E = \vec{p}^2 + g^2(\sigma^2 + \vec{\pi}^2)$ and the quark-degeneracy factor $d_n = 2N_f N_c = 12$. The mean fields have to be evaluated self-consistently from the equilibrium condition

$$\frac{\partial\Omega}{\partial\sigma} = \frac{\partial\Omega}{\partial\vec{\pi}} = 0. \tag{5}$$

In the following we restrict ourselves to vanishing pion mean fields. A nice feature of this model is that by varying the Yukawa coupling g, one finds different kinds of phase transition as illustrated in Fig. 1.

3. Semiclassical particle-field dynamics

The challenge in applying the above model to an off-equilibrium dynamical simulation of a system of particles (here quarks and antiquarks) and mean fields (representing the mesons) is that in order to reproduce the equilibrium-phase structure as depicted in 1 as the stationary limit, one has to ensure that both kinetic and chemical equilibration is possible through the introduction of the appropriate elastic collision terms for qq and $q\overline{q}$ scattering as well as quark-number changing processes such as $q\overline{q} \leftrightarrow \sigma$. In a full kinetic approach this is achieved by a set of coupled Boltzmann-Vlasov equations, which read in our case schematically (again restricting ourselves to the case of vanishing pion-mean fields)

$$\Box\sigma + \lambda(\sigma^2 - \nu^2)\sigma - f_\pi m_\pi^2 + g\left\langle\overline{\psi}\psi\right\rangle = I(\sigma \leftrightarrow \overline{q}q), \tag{6}$$

$$\left[\partial_t + \frac{p}{E_q}\cdot\vec{\nabla}_{\vec{x}} - \vec{\nabla}_{\vec{x}}E_\psi(t,\vec{x},\vec{p})\cdot\vec{\nabla}_{\vec{p}}\right] f_q(t,\vec{x},\vec{p}) = C(\psi\psi \to \psi\psi, \sigma \leftrightarrow \overline{q}q). \tag{7}$$

Here, I and C denote collision integrals contributing to the meson-mean-field and quark-phase-space distribution functions respectively.

In the following a novel scheme to Monte-Carlo simulate such a system of kinetic equations is defined, where one describes the mesons solely with a mean field and the quarks and antiquarks in terms of test particles. While the elastic-collision term is realized in a straightforward way using the corresponding cross section from the underlying linear σ model, one has to find a way to realize the interactions $q\overline{q} \leftrightarrow \sigma$ in such a scheme, while still fulfilling energy-momentum conservation and the principle of detailed balance, which are the fundamental principles constraining the off-equilibrium dynamics and ensuring the proper (Maxwell-Boltzmann) equilibrium limit.

In our recently developed model (Dynamical Simulation of a Linear Sigma Model, DSLAM) this challenge is solved as follows: In order to properly simulate the collision terms on the right-hand sides of Eqs. (6) and (7) we define a space-time grid. In each time step at each spatial

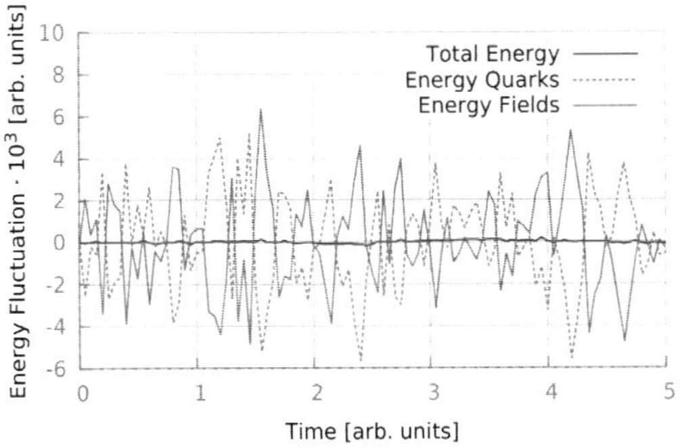

Figure 2. While the energies of the quarks and the meson field in the thermal-box simulation show anti-correlated thermal fluctuations of the order $\Delta E/E \sim 10^{-3}$ for the quarks and $\Delta E/E \sim 10^{-2}$ for the field the numerical fluctuations of the total energy amount to only $\Delta E/e \lesssim 5 \cdot 10^{-5}$.

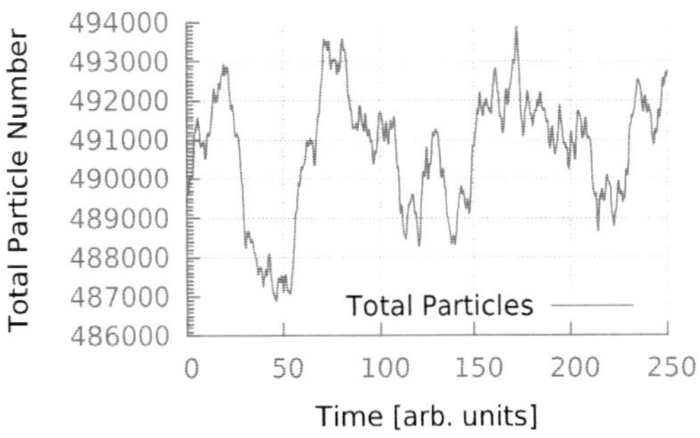

Figure 3. The total quark number in the thermal-box simulation shows fluctuations due to the dynamical pair-creation and annihilation processes, leading to an exchange of energy between the quarks and the mean field as shown in the left Fig. 2.

cell the cross section for the annihilation process $q\overline{q} \to \sigma$ is used to stochastically determine an energy-momentum transfer from the initial $q\overline{q}$ pair, located in the cell. This energy-momentum change is transferred to the σ mean field in terms of an appropriate relativistic Gaussian wave packet $\delta\sigma(t, \vec{x})$ (which simulates the gain term in I), and the $q\overline{q}$ pair is taken out of the test-particle ensemble (which simulates the corresponding loss term in C). To simulate also the appropriate decay process, $\sigma \to \overline{q}q$, we have to "particlize" the mean field locally in each spatial cell. This is done in the spirit of a course-graining procedure: First, the total energy-momentum content of the σ field within the cell in terms of the corresponding σ-field energy-momentum tensor is determined. Then one assumes a local thermal equilibrium phase-space distribution, equivalent to this energy-momentum tensor. In order to fulfill detailed balance, the temperature has to be the same as that for the corresponding procedure for the quarks and antiquarks. The temperature is related to the mean-field value which depends on the scalar quark-antiquark density. In this way a temperature can be determined. It is important to note that it is defined in the local rest frame of the heat bath and thus the σ-phase-space distribution is given by a Maxwell-Jüttner distribution $f_\sigma \propto \exp[-p_\mu u^\mu(t, \vec{x})/T]$, where u^μ is the four-velocity, given by the total field-four-momentum in the spacial cell under consideration, $u^\mu = p^\mu/E$. Now in each time-step within each spatial cell one can choose an ensemble of σ particles according to this local Maxwell-Jüttner distribution and using the corresponding $q\overline{q} \to \sigma$ decay rate to determine the gain term to C. The loss term for the mean-field equation in the collision term I is again achieved by taking the appropriate amount of energy and momentum out of the mean field in terms of a Gaussian wave packet.

In summary we have achieved a scheme which enables us to simulate the set of Boltzmann-Vlasov Eqs. (6) and (7) using test particles for the quarks and antiquarks and restricting the description of the mesons strictly to the mean-field level. The scheme by construction fulfills energy-momentum conservation through the Gaussian wave-packet description for the exchange of energy and momentum between the mean field and the test particles. At the same time also the principle of detailed balance is fulfilled, using the coarse-graining approach to locally map the field-energy-momentum distribution to a local-equilibrium Maxwell-Jüttner distribution to reinterpret the mean field as a meson phase-space distribution and using the leading-order transition rates (cross sections) of the underlying QFT linear σ model fulfilling the detailed-

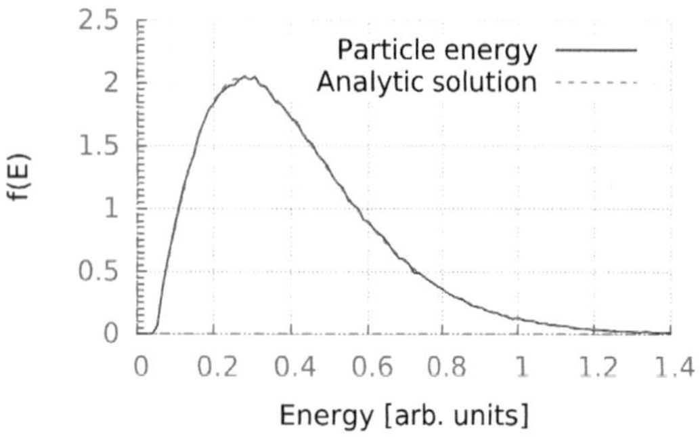

Figure 4. The distribution of the quarks' kinetic energy in a thermal-box calculation show an excellent agreement with the expected Maxwell-Boltzmann distribution.

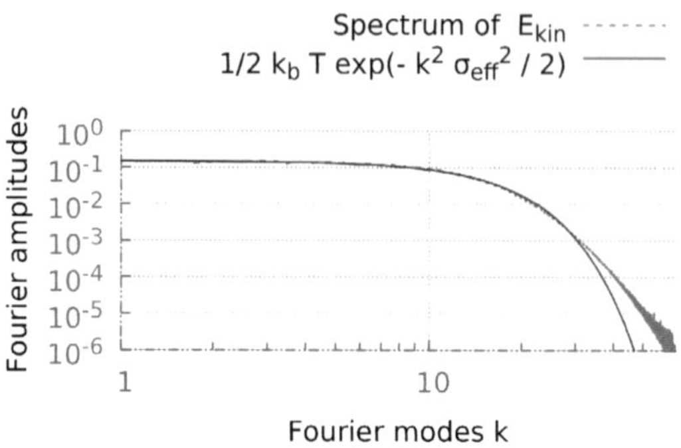

Figure 5. The distribution of the kinetic field energy to its Fourier modes. The classical "UV catastrophe", i.e., the equipartition theorem associating an average energy of $k_B T/2$ for each Fourier mode, is avoided by the use of a finite finite width $\sigma_{\rm eff}$ of the Gaussian wave packets for the transfer of energy and momentum between particles and field.

balance principle.

4. Proof of principle: "Box" calculations

As a plausibility check of the simulation method the stability of an equilibrium situation has been tested in a finite cubic box with periodic boundary conditions [14]. The stability of the energy conservation is demonstrated in Fig. 2. While the kinetic energy of the quarks and the energy of the mean field show anti-correlated thermal fluctuations the total energy stays stable within the numerical accuracy of the simulation.

The annihilation and creation processes of quark-antiquark pairs lead to thermal fluctuations in the total particle number, as shown in Fig. 3.

The energy distribution of the quarks shows the expected Maxwell-Boltzmann distribution at the expected temperature, as demonstrated in Fig. 4. Also the spectral analysis of the mean field shows that the thermal equipartition theorem for the kinetic field energy is fulfilled at small wave numbers (long wavelengths), i.e., each mode contains an average energy of $k_B T/2$. On the other hand the "UV catastrophe" must be avoided due to the finite total-energy content within a finite box. Indeed, the energy-momentum transfer between particles and the field due to the pair-annihilation and -creation processes is not strictly "local" but occur within a finite volume whose scale is fixed by the finite width, σ, of the Gaussian wave packets used as field increments to keep care of the correct energy-momentum transfer in each process. Thus the large wave numbers (short wave lengths) are effectively cut off at a scale $k_{\rm cutoff} = \sqrt{2}/\sigma$, as shown by plotting the corresponding Gaussian on top of the k-distribution of the kinetic field energy (Fig. 5).

The dynamical behavior of the field in our thermal-box simulation is illustrated in Fig. 6, showing the field distribution within the xy plane of the simulation. Starting the simulation with a uniform mean-field equilibrium value, after a short time some local blob-like disturbances have developed (left panel) due to quark-pair-annihilation processes, leading to the propagation of Gaussian wave packets on top of the still quite uniform average field value, as implemented by our concept of describing the exchange between field and particles in these processes. The right panel shows the fully developed equilibrium-thermal-field fluctuations in the long-time limit of

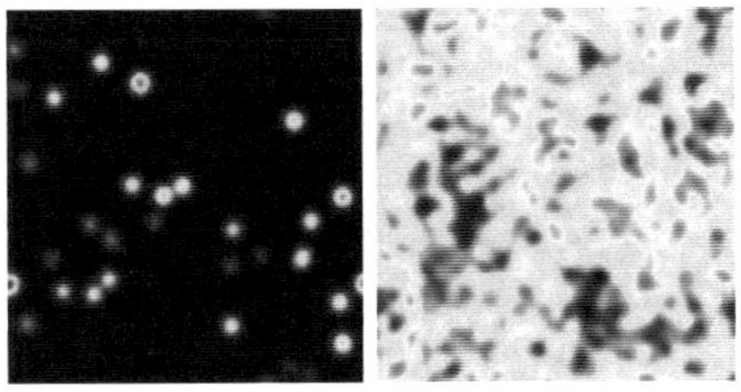

Figure 6. Starting with a uniform mean field, a short time after the start of the simulation the field is disturbed by some Gaussian wave packets due to quark-pair annihilation processes (left panel). In the long-time limit the full equilibrium-thermal-field fluctuations have developed.

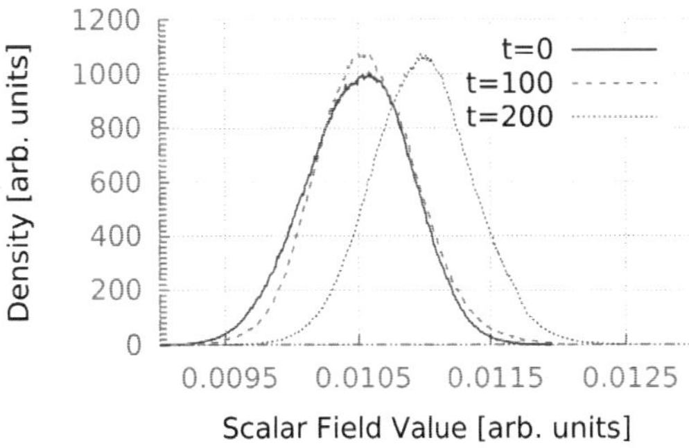

Figure 7. The field fluctuations are Gaussian distributed around an average value which can slowly drift with time due to the dynamical fluctuations of the field momentum.

the simulation. In this limit the field values are Gaussian distributed around a mean value (cf. Fig. 7), which is expected due to many random energy-momentum transfers between the field and particles. Thereby the average field value can slowly drift with time due to the thermal fluctuations of the field energy and momentum.

5. Conclusions and outlook

In this talk it was demonstrated that the dynamical description of the chiral phase transition in a simple quark-meson model is feasible with a novel Monte-Carlo-simulation technique for a corresponding coupled Boltzmann-transport equation for the meson mean-field and the quark and anti-quark phase-space distribution functions, implementing both elastic quark and anti-quark scattering as well as quark-antiquark-pair creation and annihilation processes, enabling both kinetic and chemical equilibration between particles and the mean field (mesons).

The simulation is set up in a way that both energy-momentum conservation and the principle of detailed balance are precisely realized (within the limits of achievable numerical accuracy). While the elastic quark-scattering processes are simulated with a straight-forward test-particle realization, the particle-field kinetics has demanded the development of a novel scheme.

The $q\bar{q} \to \sigma$ annihilation process is evaluated by the Monte-Carlo sampling according to the corresponding transition matrix elements from the underlying quantum-field theoretical interpretation of the linear σ model. The corresponding energy and momentum are precisely transferred to the σ-mean field in terms of an appropriate disturbance in form of a relativistic Gaussian wave packet. To obey the principle of detailed balance, also the inverse $\sigma \to q\bar{q}$ decay has to be simulated. To that purpose a coarse-graining procedure based on the local energy-momentum content of the field has been used for a "particlization" of the mean field in terms of a local Boltzmann-Jüttner equilibrium distribution, which in turn enables a Monte-Carlo sampling of the σ-meson decay according to the decay rate from the quantum-field theory picture of the model, which automatically implements detailed balance.

This scheme has now been applied to off-equilibrium situations as in a "thermal quench" in a box where the particles and fields are initialized with different temperatures and it was demonstrated that the distribution can describe a possible phase transition from the initial state to a late-time (equilibrium) state.

Last but not least also the case of "expanding fireballs", mimicking the situation of the medium created in heavy-ion collisions, is under study.

Acknowledgments

This work was partially supported by the Bundesministerium für Bildung und Forschung (BMBF Förderkennzeichen 05P12RFFTS) and by the Helmholtz International Center for FAIR (HIC for FAIR) within the framework of the LOEWE program (Landesoffensive zur Entwicklung Wissenschaftlich-Ökonomischer Exzellenz) launched by the State of Hesse. C. W. and A. M. acknowledge support by the Helmholtz Graduate School for Hadron and Ion Research (HGS-HIRe), and the Helmholtz Research School for Quark Matter Studies in Heavy Ion Collisions (HQM). Numerical computations have been performed at the Center for Scientific Computing (CSC). H. v. H. has been supported by the Deutsche Forschungsgemeinschaft (DFG) under grant number GR 1536/8-1. C.W. has been supported by BMBF under grant number 0512RFFTS.

References

[1] Friman B, Hohne C, Knoll J, Leupold S, Randrup J *et al.* 2011 *Lect.Notes Phys.* **814** pp. 980 URL `http://dx.doi.org/10.1007/978-3-642-13293-3`
[2] Philipsen O 2013 *Prog. Part. Nucl. Phys.* **70** 55–107 URL `http://dx.doi.org/10.1016/j.ppnp.2012.09.003`
[3] Nahrgang M, Leupold S, Herold C and Bleicher M 2011 *Phys. Rev. C* **84** 024912 URL `http://dx.doi.org/10.1103/PhysRevC.84.024912`
[4] Nahrgang M, Leupold S and Bleicher M 2012 *Phys. Lett. B* **711** 109–116 URL `http://dx.doi.org/10.1016/j.physletb.2012.03.059`
[5] Herold C, Nahrgang M, Mishustin I and Bleicher M 2013 *Phys. Rev. C* **87** 014907 URL `http://dx.doi.org/10.1103/PhysRevC.87.014907`
[6] Schmidt J, Meistrenko A, van Hees H, Xu Z and Greiner C 2015 *Phys. Rev. E* **91** 032125 URL `http://dx.doi.org/10.1103/PhysRevE.91.032125`
[7] Stephanov M A, Rajagopal K and Shuryak E V 1999 *Phys. Rev. D* **60** 114028 URL `http://dx.doi.org/10.1103/PhysRevD.60.114028`
[8] Schäfer B J, Pawlowski J M and Wambach J 2007 *Phys. Rev. D* **76** 074023 URL `http://dx.doi.org/10.1103/PhysRevD.76.074023`
[9] Skokov V, Friman B and Redlich K 2011 *Phys. Rev. C* **83** 054904 URL `http://dx.doi.org/10.1103/PhysRevC.83.054904`
[10] Skokov V, Stokic B, Friman B and Redlich K 2010 *Phys. Rev. C* **82** 015206 URL `http://dx.doi.org/10.1103/PhysRevC.82.015206`
[11] Braun-Munzinger P, Friman B, Karsch F, Redlich K and Skokov V 2012 *Nucl. Phys. A* **880** 48–64 URL `http://dx.doi.org/10.1016/j.nuclphysa.2012.02.010`
[12] Schaefer B J 2012 *Phys. Atom. Nucl.* **75** 741–743
[13] Morita K, Skokov V, Friman B and Redlich K 2014 *Eur. Phys. J. C* **74** 2706 URL `http://dx.doi.org/10.1140/epjc/s10052-013-2706-1`
[14] Wesp C, van Hees H, Meistrenko A and Greiner C 2015 *Phys. Rev. E* **91** 043302 URL `http://dx.doi.org/10.1103/PhysRevE.91.043302`
[15] Gell-Mann M and Lévy M 1960 *Nuovo Cim.* **16** 705 URL `http://dx.doi.org/10.1007/BF02859738`

Overview and highlights of bulk correlations at STAR

Daniel McDonald

E-mail: dlmcdona@central.uh.edu

Abstract. The first phase of the Beam Energy Scan (BES) program at the Relativistic Heavy Ion Collider (RHIC) was completed during the years 2010, 2011 and 2014, with Au+Au collisions at center-of-mass energies ($\sqrt{s_{\rm NN}}$) of 7.7, 11.5, 14.5, 19.6, 27, and 39 GeV. The BES has three distinct goals: search for the turning off of the signatures of the Quark Gluon Plasma (QGP), search for the first-order phase transition, and search for the critical point of nuclear matter. We report several interesting results on each of these fronts. In addition, we report the first bulk correlation results from U+U collisions at STAR from a three week exploratory run in 2012.

1. Introduction

Heavy-ion collisions at top energies at RHIC produce a new phase of nuclear matter known as the Quark Gluon Plasma (QGP), where quarks and gluons are no longer confined within each individual nucleon but are asymptotically free in a region where net-baryon density is vanishing [1]. Little is known about the phase diagram of quark (Quantum Chromodynamics, QCD) matter. The scientific goal of the BES is to explore this phase diagram, significantly increasing the scientific community's knowledge in a very nascent field of study.

Figure 1 shows a schematic of the QCD phase diagram, with temperature on the y-axis and baryochemical potential, $\mu_{\rm B}$, on the x-axis. In the upper right is the QGP phase, and in the lower left is the Hadronic Gas (HG) phase. Theoretical trajectories at various BES energies are shown in yellow. At very low $\mu_{\rm B}$, lattice QCD calculations predict a smooth crossover transition [2, 3, 4]. Meanwhile, at higher values of baryochemical potential, QCD-based models predict that there will be a first-order phase transition between the QGP and HG [5, 6]. At the low-$\mu_{\rm B}$ end of the first-order phase transition line is a posited critical point.

The BES has three distinct goals, searching for: 1) the turning off of the signatures of the QGP, 2) the first-order phase transition, and 3) the critical point. Table 1 lists details of the BES data sets. In addition, RHIC successfully collided uranium nuclei in a three week exploratory run in 2012, collecting 360 million minimum bias events, as well as 13 million central top 1 percent events according to the Zero-Degree Calorimeter (ZDC). This 193 GeV U+U dataset aims to utilize uranium's unique prolate shape to explore the effects of different collision geometries.

2. Search for the turning off of QGP signatures

2.1. Nuclear modification factor, $R_{\rm CP}$

At top RHIC energies, we observed the energy loss of high-energy partons and the associated suppression in the production of mesons with high transverse momentum ($p_{\rm T}$) in Au+Au collisions [7]. This nuclear modification factor is denoted by $R_{\rm CP}$, which is the ratio of the number of particles in central A+A collisions to the number of particles in peripheral A+A

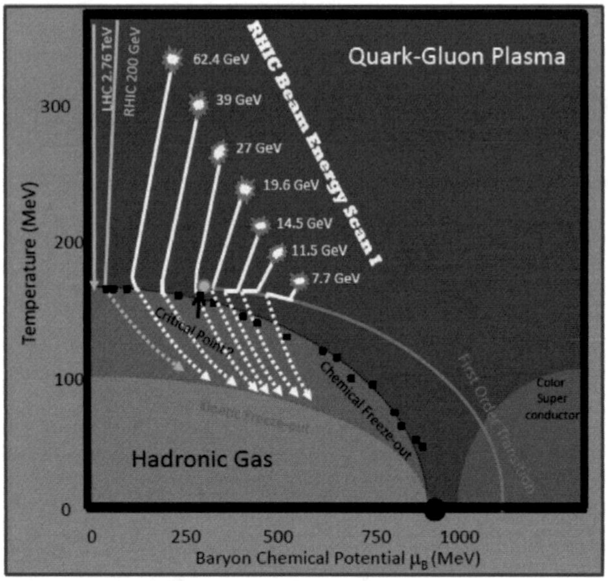

Figure 1. Schematic of the QCD phase diagram with trajectories (yellow) of the RHIC BES program. Trajectories are for illustrative purposes only.

Table 1. Overview of the Beam Energy Scan data sets

Energy (GeV)	Events (M)	Time (Weeks)
200	350	11
62.4	67	1.5
39	130	2
27	70	1
19.6	36	1.5
14.5	20	3
11.5	12	2
7.7	4	4

collisions, scaled by the number of binary collisions.

Figure 2 shows R_{CP} versus p_{T} at seven different beam energies at STAR. Suppression at high p_{T} is indicative of QGP formation. No suppression is seen below 39 GeV, but further statistics are needed to extend the p_{T} reach to lower beam energies.

2.2. Elliptic flow

The second-order Fourier coefficient of the azimuthal particle distribution relative to the reaction plane (RP) is given by the elliptic flow v_2, where $v_2 = \langle \cos[2(\varphi - \Psi_{\mathrm{RP}})] \rangle$, and Ψ_{RP} is the azimuth of the reaction plane and φ is the azimuthal angle.

Figure 3 shows v_2 scaled by the number of constituent quarks (n_Q) as a function of n_Q-scaled m_{T}-m for identified hadrons for 0-80% Au+Au collisions as published by the STAR collaboration at BES energies [8]. Scaling by the number of constituent quarks is thought to be indicative of partonic behavior. In Fig. 3, n_Q scaling holds for most particles to within ~10%. The ϕ meson may not follow the trend of the other identified hadrons at 7.7 and 11.5 GeV, but more data are needed since the associated errors are quite large.

Journal of Physics: Conference Series 636 (2015) 012008 doi:10.1088/1742-6596/636/1/012008

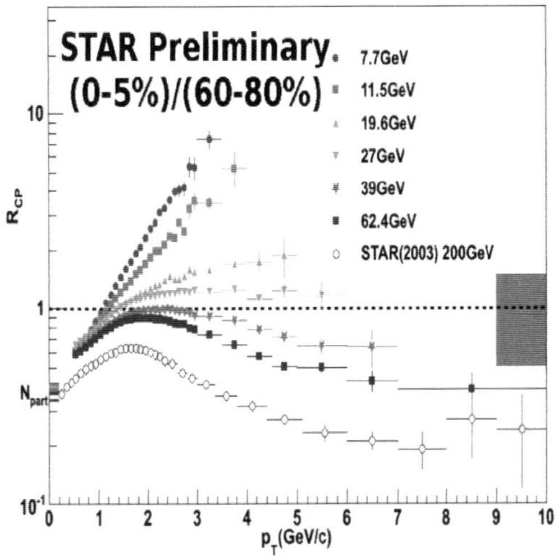

Figure 2. Nuclear modification factor, $R_{\rm CP}$, as a function of transverse momentum, $p_{\rm T}$, at different beam energies.

When one looks at the difference in v_2 between particles and their associated antiparticles for 0-80% Au+Au collisions at STAR [9], the particle and antiparticle v_2 agree well at 200 GeV, and the difference between the particle and antiparticle v_2 increases as the beam energy decreases. This difference is almost linearly proportional to $\mu_{\rm B}$, implying a correlation between the v_2 difference of particles and antiparticles and the net-baryon density.

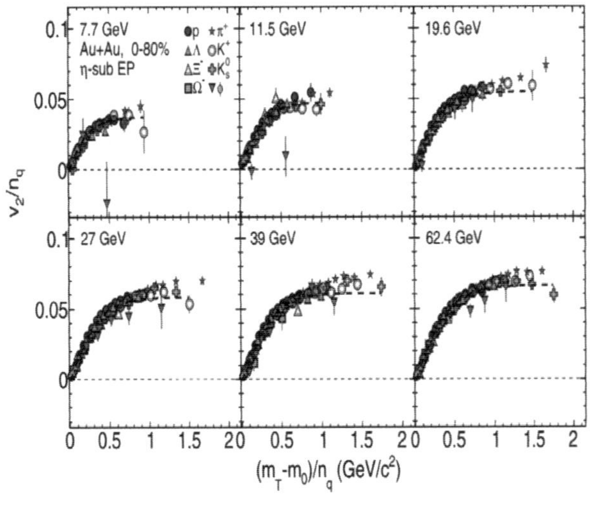

Figure 3. v_2 scaled by the number of constituent quarks (n_Q) for identified hadrons as a function of the n_Q-scaled m_T-m for 0-80% Au+Au collisions from 7.7-62.4 GeV at STAR [8].

2.3. Chiral Magnetic Effect
The Chiral Magnetic Effect (CME) is a phenomenon where charges separate as a result of the large magnetic fields that form in non-central collisions and the intrinsic QCD topological structure [10]. The CME requires the system to be deconfined; therefore, the turnoff at lower beam energies of the CME would be a turn off of a QGP signature. Figure 4 shows the three-point correlator, γ, versus centrality in Au+Au collisions at STAR from 7.7-62.4 GeV for opposite charge (circles) and same charge (stars) [12]. The value of γ, a means of experimentally measuring the charge separation, could be sensitive to the CME [13]. One can see that as the energy decreases, the separation decreases and is consistent with zero at 7.7 GeV.
One issue with the CME observalbles is the potentially large flow-related background. One way to explore the sensitivity of charge separation to the CME is to collide U+U, since the prolate shape of uranium allows one to create systems with finite elliptic flow but a small magnetic field. Figure 5 shows the charge separation signal as a function of v_2 in U+U collisions at 193 GeV [14]. The 0-1% most central events are indicated by the blue circle near a y-axis value of 0. As one can see, at a finite value of v_2, the value is consistent with zero in the most central events is consistent with the expectation of CME with a small magnetic field and minimal flow-related background in those events.

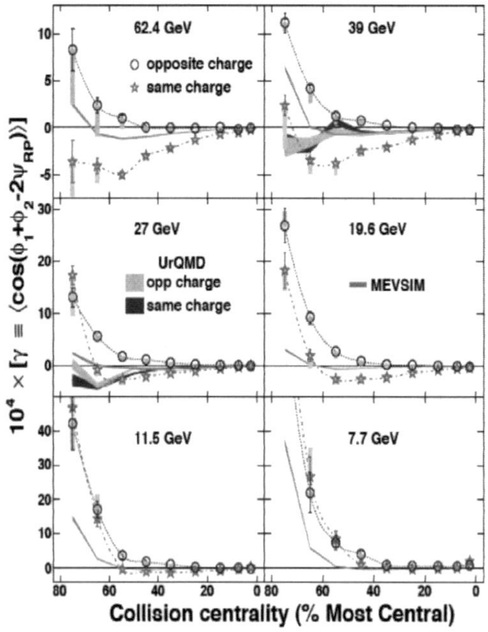

Figure 4. The three-point correlator (γ) as a function centrality in Au+Au collisions from 7.7-62.4 GeV [12] for opposite charge (circles) and same charge (stars).

3. Search for the first-order phase transition

3.1. Azimuthal anisotropy, v_n
A first-order phase transition would be indicated by a region with the lowest compressibility. According to hydrodynamic calculations that contain a first-order phase transition, the value of v_1 of net-baryons can be used as a probe of the phase transition [15]. A non-monotonic variation

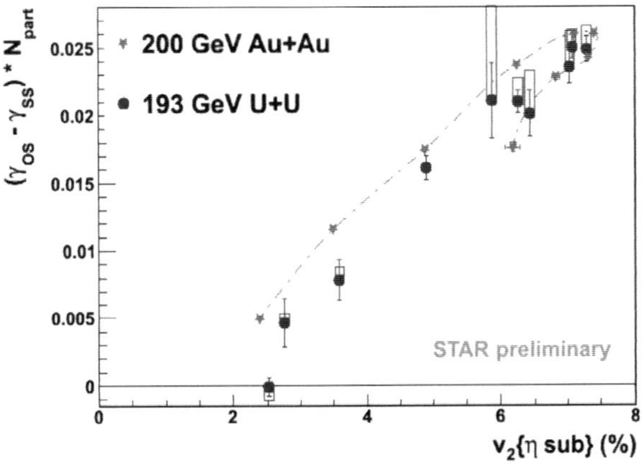

Figure 5. $\gamma_{OS}-\gamma_{SS}$*$N_{\rm part}$ versus v_2 for Au+Au collisions at 200 GeV (red stars) and U+U collisions at 193 GeV (blue circles) [14].

of the slope of this directed flow would signal a softening of the equation of state.
Figure 6 shows the slope of directed flow (dv_1/dy) near mid-rapidity of antiprotons (upper), protons (mid), and net protons (lower) for 10-40% centrality versus beam energy in Au+Au collisions from 7.7 to 200 GeV at STAR [16]. The proton values change sign between 7.7 and 11.5 GeV, have a minimum between 11.5 and 19.6 GeV, and then increase but are still negative at 200 GeV. The net-proton result shows a similar local minimum between 11.5 and 19.6 GeV that could be a signal of the softening of the equation of state.
When searching for a softening of the equation of state, one looks for observables that are proportional to pressure, such as the v_n coefficients. In addition to v_1, recent work suggests that v_3 could be more promising. Some model predictions suggest that a hydro phase is not necessary to create v_2; however, v_3 is more sensitive to the hydro phase of the fluid created in heavy-ion collisions [17]. Recent STAR results show that while multiplicity increases with energy, v_3 does not increase until ~20 GeV, suggesting that this observable warrants further investigation.

3.2. Femtoscopy
The study of two-pion interferometry is typically referred to as the study of HBT (Hanbury Brown and Twiss). HBT has proven to be a precise tool for measuring the pressure anisotropy and lifetime of the system [19]. Since a first-order phase transition involves a mixed-phase regime with pressure gradients of zero, a measurement of the freezeout shape should show non-monotonic behavior near the first-order phase transition [20].
Pion azimuthal HBT allows the study of the coordinate-space almond shape after expansion, given by the kinetic freezeout eccentricity $\epsilon_{\rm F}$. Early measurements of $\epsilon_{\rm F}$ hinted at possible non-monotonic behavior at CERES [21]. Figure 7 shows $\epsilon_{\rm F}$ versus beam energy, where the STAR datapoints are shown as colored stars [19]. Data from CERES are shown as a plus sign and E895 [22] are shown as squares [23]. Even when changing the rapidity range, STAR does not see any non-monotonic behavior and $\epsilon_{\rm F}$ agrees well with the trend of the Boltzmann transport model, UrQMD [24].

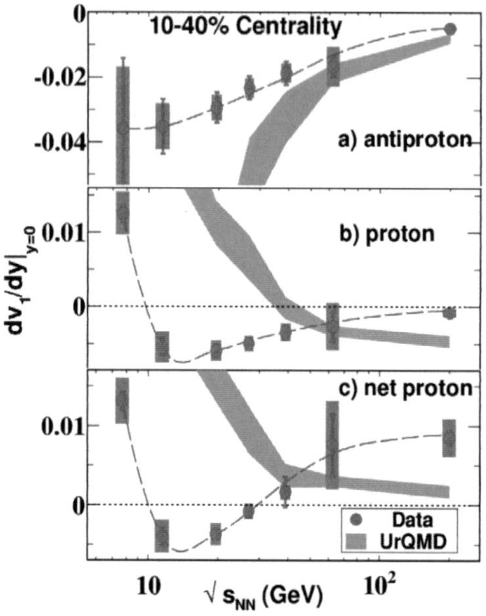

Figure 6. The slope of directed flow (dv_1/dy) near mid-rapidity versus beam energy for mid-central (10-40%) Au+Au collisions at STAR, as well as UrQMD calculations [16].

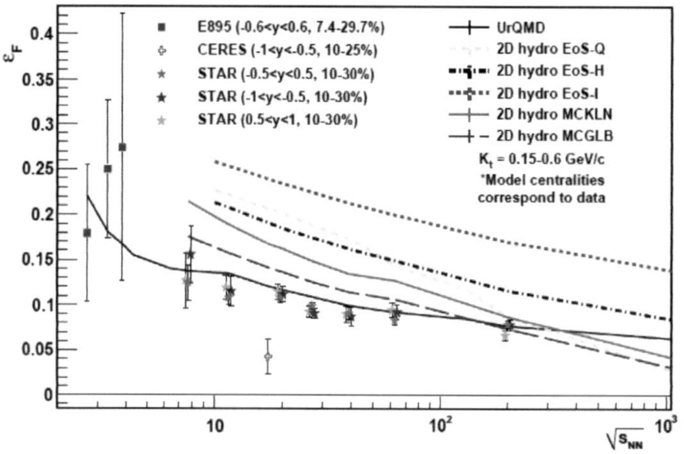

Figure 7. The beam-energy dependence of the kinetic freezeout eccentricity ϵ_F for Au+Au collisions at STAR, E895 and Pb+Au collisions at CERES. Three different rapidity ranges (forward, mid, backward) are shown for the STAR energies as colored stars (green, red, blue, respectively) [19].

4. Search for the critical point

4.1. Higher-order net-particle moments

Section 3 discusses efforts to locate the first-order phase transition of QCD matter. If there is a first-order phase transition, then there should be a critical point at the end of the first-order phase transition line before the transition becomes a crossover [25]. The susceptibilities of conserved quantities (electric charge, strangeness, and baryon-number), which are directly related to the experimentally measurable cumulants of the multiplicity distributions, should have distinctly different values in the QGP and hadronic phases [26]. In addition, phenomenological models predict that the moments of multiplicity distributions sensitive to the correlation length will diverge near the CP, and the magnitude of the non-monotonic behavior should increase as the order of the cumulant is increased [27].

Net-proton was used as a proxy of baryon number [28], and net-charge was used as a proxy of electric charge. Figure 8 shows the moment products $S\sigma$ (upper) and $\kappa\sigma^2$ (middle) for net-protons in 0-5% (circles) and 70-80% (squares) collisions at STAR [29]. The lower frame shows the values of the moment product $S\sigma$ scaled by the expectation of a Skellam distribution. The Skellam expectations are shown as solid lines. A possible non-monotonic variation of $\kappa\sigma^2$ relative to unity is not excluded by the existing STAR data. The STAR data does agree well with the expectation from independent production, which breaks intra-event correlations between the numbers of protons and antiprotons. Higher statistics will be necessary to further elucidate results, yielding exciting possibilities for a second phase of the Beam Energy Scan.

The moment products σ^2/M (top), $S\sigma$ (middle), and $\kappa\sigma^2$ (bottom) for net-charge show no evidence of non-monotonic behavior within statistical uncertainties [30].

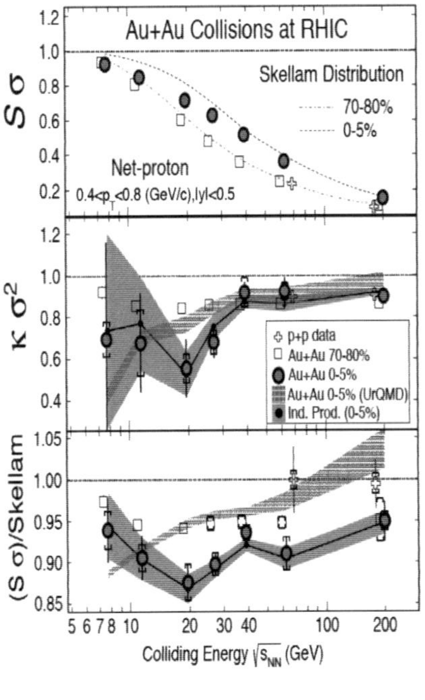

Figure 8. The beam-energy dependence of the net-proton moment products $S\sigma$ (top) and $\kappa\sigma^2$ (middle) at two different centralities. The hatched band shows the expectation from UrQMD and the shaded band shows the expectation assuming independent proton and antiproton production. The bottom frame shows the value of $S\sigma$ scaled by the Skellam expectation shown in the upper frame (lines).

5. Dileptons and the chiral transition

Dileptons are ideal probes of heavy-ion collisions since they are emitted throughout the evolution of the system and are inert to the strong force [31]. In the low invariant mass region

(LMR), where M_{ee}<1.1 GeV/c^2, the spectral shape of the ρ meson is sensitive to in-medium modifications, so it could be sensitive to the chiral phase transition [32]. One could also potentially study QGP thermal radiation via the dilepton measurements in the intermediate invariant mass region (IMR) of 1.1<M_{ee}<3.1 GeV/c^2 [33].

In recent STAR results, there is no strong beam-energy dependence of the LMR excess. Further comparison with SPS would be possible with energies below 19.6 GeV. The excess could spike near the CP because of anomalous increases in the lifetime of the system [34]. Statistics are insufficient for meaningful interpretation of the results in the IMR region, but this observable should be pursued in a second phase of the Beam Energy Scan.

6. Summary and Outlook

We have shown a wide array of interesting results from the first phase of the BES at RHIC. When looking for the turning off of QGP signatures, signs of deconfinement remain down to at least ~27 GeV. Some observables, such as v_1, point to the potential softening of the equation of state/first-order phase transition in the region of ~20 to 40 GeV. Further statistics at the lowest BES beam energies are needed before reaching conclusions about the critical point. While there may be some tantalizing hints of interesting phenomena in the RHIC BES, firmer conclusions could be reached with a second phase of the BES [18]. Early results of U+U collisions at STAR already hint at some interesting physics. For example, in 0-1% central collisions the charge separation signal disappears while v_2 remains finite, consistent with a picture of the CME.

[1] J. Adams et al. [STAR Collaboration], Nucl. Phys. A **757**, 28 (2005).
[2] Y. Aoki, S. Borsanyi, S. Durr, Z. Fodor, S. D. Katz, S. Krieg and K. K. Szabo, JHEP **0906**, 088 (2009).
[3] S. Borsanyi et al., [Wuppertal-Budapest Collaboration], JHEP **1009**, 073 (2010).
[4] A. Bazavov et al., Phys. Rev. D **85**, 054503 (2012).
[5] S. Ejiri, Phys. Rev. D **78**, 074507 (2008).
[6] E. S. Bowman and J. I. Kapusta, Phys. Rev. C **79**, 015202 (2009).
[7] J. Adams et al., [STAR Collaboration], Phys. Rev. Lett. **91**, 172302 (2003).
[8] L. Adamczyk et al., [STAR Collaboration], Phys. Rev. C **88**, 014902 (2013).
[9] L. Adamczyk et al., [STAR Collaboration], Phys. Rev. Lett. **110**, 142301 (2013).
[10] D. Kharzeev, Phys. Lett. B **633**, 260 (2006).
[11] B. Abelev et al., [ALICE Collaboration], Phys. Rev. Lett. **110**, 012301 (2013).
[12] L. Adamczyk et al., [STAR Collaboration], Phys. Rev. Lett. **113**, 052302 (2014).
[13] S. Voloshin, Phys. Rev. C **70**, 057901 (2004).
[14] G. Wang for the STAR Collaboration, Nucl. Phys A **904**, 248c (2013).
[15] H. Stoecker, Nucl. Phys. A **750**, 121 (2005).
[16] L. Adamczyk et al., [STAR Collaboration], Phys. Rev. Lett. **112**, 162301 (2014).
[17] J. Auvinen, H. Peterson, Phys. Rev. C **88**, 064908 (2013).
[18] STAR Collaboration. Studying the Phase Diagram of QCD Mattter at RHIC. SN0598 (2014).
[19] L. Adamczyk et al., [STAR Collaboration], (2014), arXiv:1403.4972v1.
[20] P. F. Kolb and U. W. Heinz (2003), arXiv:nucl-th/0305084.
[21] M. A. Lisa et al., New J. Phys. **13**, 065006 (2011).
[22] M. A. Lisa et al. [E895 Collaboration], Phys. Lett. B **496**, 1 (2000).
[23] D. Adamova et al. [CERES Collaboration], Phys. Rev. C **78**, 064901 (2008).
[24] H. Petersen, M. Bleicher, S. A. Bass, and H. Stocker (2008), arXiv:0805.0567.
[25] Y. Aoki, G. Endrodi, Z. Fodor, S. D. Katz and K. K. Szabo, Nature, **443**, 675 (2006).
[26] M. Cheng et al., Phys. Rev. D **79**, 074505 (2009).
[27] C. Athanasiou et al., Phys. Rev. D **82**, 074008 (2010).
[28] Y. Hatta and M. A. Stephanov, Phys. Rev. Lett. **91**, 102003 (2003).
[29] L. Adamczyk, et al., [STAR Collaboration], Phys. Rev. Lett. **112**, 032302 (2014).
[30] L. Adamczyk, et al., [STAR Collaboration], Phys. Rev. Lett. **113**, 092301 (2014).
[31] J. Alam, S. Sarkar, P. Roy, T. Hatsuda and B. Sinha, Annals Phys. **286**, 159 (2001).
[32] G. E. Brown and M. Rho, Phys. Rept. **269** (1996) 333.
[33] E. V. Shuryak, Phys. Rept. **61** (1980) 71.
[34] R. Rapp, Adv. High Energy Phys. 2013, 148253 (2013).

ALICE summary of light flavour results at intermediate and high p_{T}

Tuva Richert, on behalf of the ALICE collaboration

Lund University, Department of Physics, Div. of Particle Physics, Box 118, SE-221 00 Lund

E-mail: tuva.richert@hep.lu.se

Abstract. The ALICE experiment has unique capabilities for particle identification at mid rapidity over a wide range of transverse momenta (p_{T}), making it an ideal tool for comprehensive measurements of hadrons such as charged pions, kaons, and protons as well as Λ, $\mathrm{K}_{\mathrm{s}}^{0}$ and ϕ. The transverse momentum distributions and nuclear modification factors, R_{pPb} and R_{PbPb}, of these hadrons measured in p–Pb and Pb–Pb collisions are presented. Baryon-to-meson ratios exhibit a multiplicity-dependent enhancement at intermediate transverse momenta for both p–Pb and Pb–Pb collisions, while no significant dynamics is observed in the ratios at larger transverse momenta. Finally, measurements of identified particle ratios in association with high-p_{T} particles as well as within reconstructed jets are presented.

1. Introduction

During LHC Run-1 the ALICE detector has recorded pp, p–Pb, and Pb–Pb collisions at different center of mass energies. Heavy-ion collisions at ultra relativistic energies are expected to produce a QCD matter where the quarks and gluons are in a deconfined state. Measurements of the production of hadrons in Pb–Pb collisions at intermediate and high p_{T}, relative to pp collisions, provide information about the dynamics of this matter. In the context of light-flavour production, the focus for the Pb–Pb results is on parton energy loss — expected to lead to a modification of energetic jets (jet quenching) — and possibly modified fragmentation due to the hot and dense QCD medium.

The excellent tracking and particle identification capabilities of the ALICE experiment, in particular its large time projection chamber, makes it possible to investigate the spectra of baryons and mesons. The results are presented in terms of particle ratios and nuclear modification factors, R_{AA} and R_{pA}.

2. R_{AA} and R_{pA} for charged hadrons

The nuclear modification factor is defined as the ratio of the particle yield in Pb–Pb to that in pp collisions scaled by the number of binary nucleon-nucleon collisions

$$R_{\mathrm{AA}}(p_{\mathrm{T}}) = \frac{\mathrm{d}^2 N_{\mathrm{AA}}/\mathrm{d}\eta \mathrm{d}p_{\mathrm{T}}}{\langle T_{\mathrm{AA}}\rangle \mathrm{d}^2\sigma_{\mathrm{pp}}/\mathrm{d}\eta \mathrm{d}p_{\mathrm{T}}} \tag{1}$$

where $\mathrm{d}^2 N_{\mathrm{AA}}/\mathrm{d}\eta \mathrm{d}p_{\mathrm{T}}$ is the differential particle yield in Pb–Pb collisions, $\mathrm{d}^2\sigma_{\mathrm{pp}}/\mathrm{d}\eta \mathrm{d}p_{\mathrm{T}}$ is the invariant cross section for particle production in inelastic pp collisions, and $\langle T_{\mathrm{AA}}\rangle$ is the average nuclear overlap function [1]. In the absence of nuclear modifications $R_{\rm AA}$ is unity for hard processes which are expected to exhibit binary collision scaling.

The nuclear modification factor presented in Fig. 1, shows that the shape of the invariant yield for peripheral Pb–Pb collisions at $\sqrt{s_{\rm NN}} = 2.76$ TeV is similar to those observed in pp collisions due to the flatness of the $R_{\rm PbPb}$, while a strong suppression of charged hadron production at high $p_{\rm T}$ is observed for central collisions. To establish whether the initial state of the colliding nuclei plays a role in the observed suppression, also the nuclear modification factor in p–Pb for charged particles is shown in the same figure. $R_{\rm pPb}$ is consistent with unity for $p_{\rm T} > 2$ GeV/c, and hence the suppression in Pb–Pb collisions is not due to any initial state effects, but to final nuclear matter effects, such as jet quenching in the hot QCD medium.

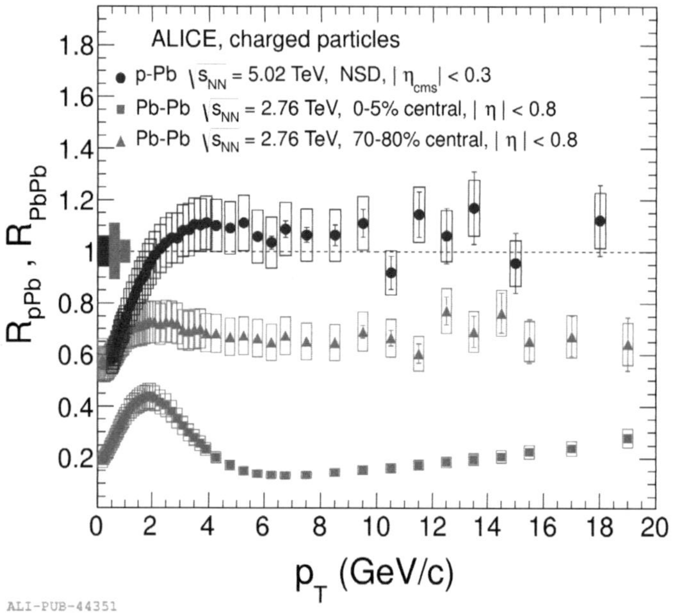

Figure 1. Nuclear modification factors, $R_{\rm PbPb}$ and $R_{\rm pPb}$, of charged particles as a function of transverse momentum for 0-5% central and 70-80% peripheral Pb–Pb at $\sqrt{s_{\rm NN}} = 2.76$ TeV collisions, and central p–Pb at $\sqrt{s_{\rm NN}} = 5.02$ TeV collisions [1].

3. R$_{\rm AA}$ and R$_{\rm pA}$ for identified hadrons

When constructing $R_{\rm AA}$ for identified light flavour hadrons, we see in Fig. 2 that, within systematic and statistical uncertainties, they are equally suppressed at $p_{\rm T} > 10$ GeV/c. The large suppression is a sign of considerable energy loss, and the $R_{\rm pA}$ seen in Fig. 3 establishes that this energy loss is predominantly due to the medium and not caused by initial state effects.

For the intermediate $p_{\rm T}$ range, the protons are less suppressed than the kaons and pions, and a mass ordering is present in the suppression pattern. While the proton and ϕ modification factors exhibit rather distinct features, the ϕ/p ratio in central Pb–Pb is observed to be approximately constant as a function of $p_{\rm T}$, as discussed in Sec. 4, indicating that the differences in the $R_{\rm AA}$ can be attributed to different pp spectra shapes.

Looking at identified particles at intermediate $p_{\rm T}$ in p–Pb collisions instead, we see in Fig. 3 an enhancement of Ξ and p, while K and π are consistent with $T_{\rm AA}$-scaled pp values. Furthermore, there is a mass ordering among π, K, p, Ξ, but the ϕ does not fit into this pattern.

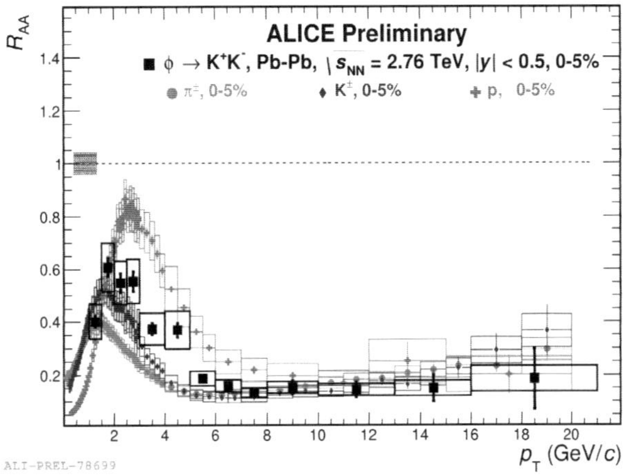

Figure 2. Nuclear modification factor, R_{AA}, of charged π, K, p, and ϕ up to 20 GeV/c in 0-5% central $\sqrt{s_{\mathrm{NN}}} = 2.76\,\mathrm{TeV}$ Pb–Pb collisions.

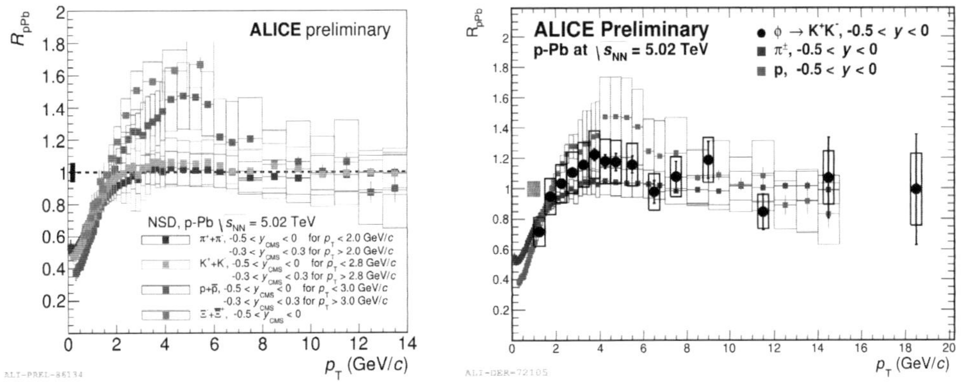

Figure 3. Nuclear modification factor, R_{pA}, for charged π, K, p, Ξ (left) and ϕ (right) in central $\sqrt{s_{\mathrm{NN}}} = 5.02\,\mathrm{TeV}$ p–Pb collisions.

4. $\Lambda/\mathrm{K^0_s}$, p/π and K/π ratios in Pb–Pb collisions

In Pb–Pb collisions, both $\Lambda/\mathrm{K^0_s}$ and p/π (Fig. 4 [2], [3]) in central and peripheral collisions are consistent with pp for $p_{\mathrm{T}} > 8$ GeV/c, indicating that the processes are dominated by vacuum-like fragmentation.

Looking in the intermediate p_{T} range for $\Lambda/\mathrm{K^0_s}$, an enhancement is visible towards more central collisions, see Fig. 4, and a shift of the maximum position towards higher p_{T} is observed: in the most peripheral collisions (60-80% centrality) there is a maximum of about 0.55 at $p_{\mathrm{T}} \sim 2$ GeV/c, while the maximum value of the ratio for the most central collisions (0-5% centrality) is about 1.6 at $p_{\mathrm{T}} \sim 3.2$ GeV/c. This shift is consistent with an increasing radial flow towards more central collisions. The magnitude of these maxima increases by almost a factor of three between most peripheral and most central Pb–Pb collision. A hydrodynamical model such as VISH2+1 [4] is able to describe the rise at low p_{T}. At higher p_{T}, models with modified fragmentation (EPOS [5], [6]) and coalescence of quarks (Recombination [7]) describe the shape qualitatively well, but overestimate the enhancement [2].

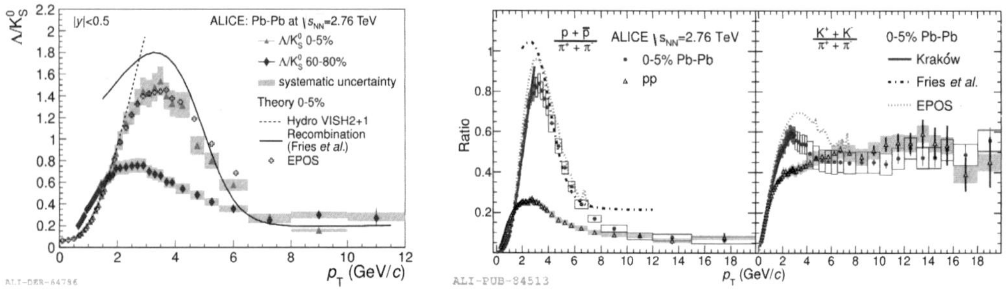

Figure 4. Left: $\Lambda/\mathrm{K}^0_\mathrm{S}$ particle ratio as a function of p_T in central (0-5%) and peripheral (60-80%) Pb–Pb collisions at $\sqrt{s_\mathrm{NN}} = 2.76\,\mathrm{TeV}$ compared to models in central (0-5%) events [2]. Right: p/π and K/π ratio as a function of p_T in central (0-5%) Pb–Pb collisions at $\sqrt{s_\mathrm{NN}} = 2.76\,\mathrm{TeV}$ compared to pp collisions at $\sqrt{s} = 7\,\mathrm{TeV}$ and models [3].

Figure 4 also shows the p/π and K/π ratio up to $p_\mathrm{T} = 20$ GeV/c for central events, both presenting an enhancement at intermediate p_T, with the peak at $p_\mathrm{T} = 3$ GeV/c. However, the baryon-to-meson ratio p/π presents a much more pronounced increase, reaching a value of about 0.9 at $p_\mathrm{T} = 3$ GeV/c, compared to the two-meson ratio K/π. As for the $\Lambda/\mathrm{K}^0_\mathrm{S}$ case, the ratios are in good agreement with hydrodynamical calculations (Krakow [4]) for $p_\mathrm{T} < 2$ GeV/c, indicating that the rise of the peak can be described by the mass ordering induced by radial flow. At intermediate p_T, around the maxima and up to p_T ~8 GeV/c, the data are qualitatively described by the recombination model by Fries *et al.* [7], and the EPOS model [5], [6], but these models also overestimate the maximum values.

5. p/ϕ in Pb–Pb collisions
To further investigate the main driving parameter in the spectral shape, we study a baryon-to-meson ratio in which the baryon and meson are of similar mass, namely the p/ϕ ratio. In Fig. 5 [8] the p/ϕ ratio is shown as a function of p_T, and it is observed that in *central* Pb–Pb collisions there is a very small difference in their p_T distributions, i.e. no baryon-meson difference is present. This indicates that the hadron mass determines the spectral shape.

6. Particle ratios in p–Pb collisions
Interestingly, the $\Lambda/\mathrm{K}^0_\mathrm{S}$ and p/π in p–Pb collisions show the same qualitative behavior as in Pb–Pb collisions: a multiplicity dependent baryon-to-meson enhancement at intermediate p_T ~3 GeV/c is seen in Fig. 6 [9] for two different multiplicity event classes. The results show that p–Pb presents features that are similar to Pb–Pb phenomenology, even though the magnitude of the enhancement in p–Pb is significantly different to the one observed in Pb–Pb. The maximum of the p/π ratio reaches 0.8 in central Pb–Pb collisions, but only 0.4 in the highest multiplicity p–Pb events, and the $\Lambda/\mathrm{K}^0_\mathrm{S}$ maximum in central Pb–Pb is 1.5, while it is 0.8 in corresponding p–Pb collisions. The highest multiplicity bin in p–Pb collisions exhibits ratios of p/π and $\Lambda/\mathrm{K}^0_\mathrm{S}$ which have maxima close to the corresponding ratios in the 60-70% centrality bin in Pb–Pb collisions, but differ somewhat in shape at lower p_T [9].

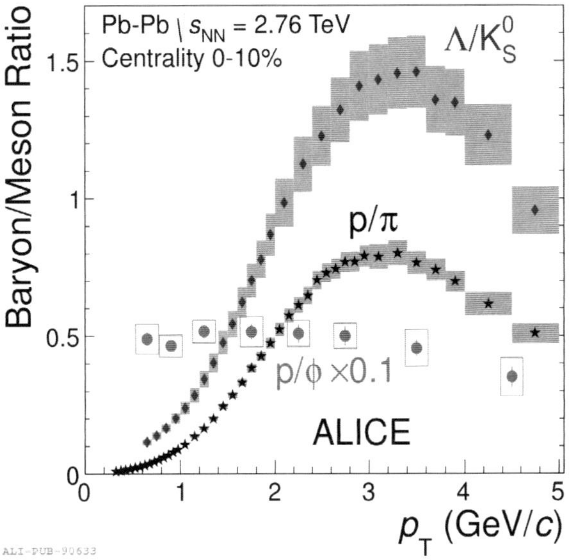

Figure 5. Ratio p/ϕ (scaled with 0.1 for visibility) as a function of p_{T} up to 5 GeV/c for 0-10% central Pb–Pb collisions at $\sqrt{s_{\mathrm{NN}}} = 2.76\,\mathrm{TeV}$, compared to $\Lambda/\mathrm{K}^0_{\mathrm{s}}$ and p/π in the same centrality bin. Figure taken from [8].

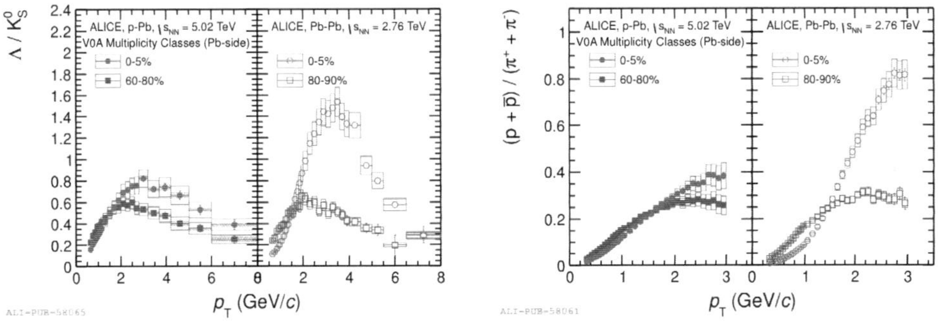

Figure 6. $\Lambda/\mathrm{K}^0_{\mathrm{s}}$ (left) and p/π (right) in $\sqrt{s_{\mathrm{NN}}} = 5.02\,\mathrm{TeV}$ p–Pb collisions (left panel) as a function of p_{T} in one central (0-5%) and one peripheral (60-80%) multiplicity bin. The ratios are compared to results in Pb–Pb collisions (right panels) where central multiplicity bin is 0-5%, and the peripheral multiplicity bin is 80-90%. Figures from [9].

7. The origin of the enhancement

One can investigate wether the origin of this enhancement is due to parton fragmentation (hard) or collective effects (soft) by a two-particle correlation study, where the particles produced in the underlying events, the *bulk*, are separated from those which are associated with a high-p_{T} trigger particle, representing a jet-like environment, or the *peak* region. The peak is defined as a region around $(\Delta\eta, \Delta\varphi) = (0, 0)$, and the bulk region around $(\Delta\eta, \Delta\varphi) = (\pm 1, 0)$ where one, due to long range (in rapidity) azimuthal correlations, expects the flow structure of the underlying event to be the same as under the peak. To study the jet contribution, the bulk is subtracted from the peak region (in Fig. 7: "Peak-Bulk"). In Fig. 7 (top) [10], the p/π ratios in central

Pb–Pb events are presented for bulk and for peak-bulk event selections, and it is seen that the enhancement is a bulk effect and nt present in jet events.

In the p–Pb study, charged particle jets are reconstructed on an event-by-event basis using an anti-k_T algorithm with resolution parameter $R = 0.2$, 0.3, or 0.4 and requiring one charged track with $p_T > 10$ GeV/c. The Λ and K^0_S yields are measured within the jet cone and corrected for the underlying event before the ratio is taken. When the ratio is compared to the inclusive ratio, the same conclusion as for the p/π ratio can be drawn: that the baryon-to-meson enhancement originates from the bulk, and is not present in the jet structure.

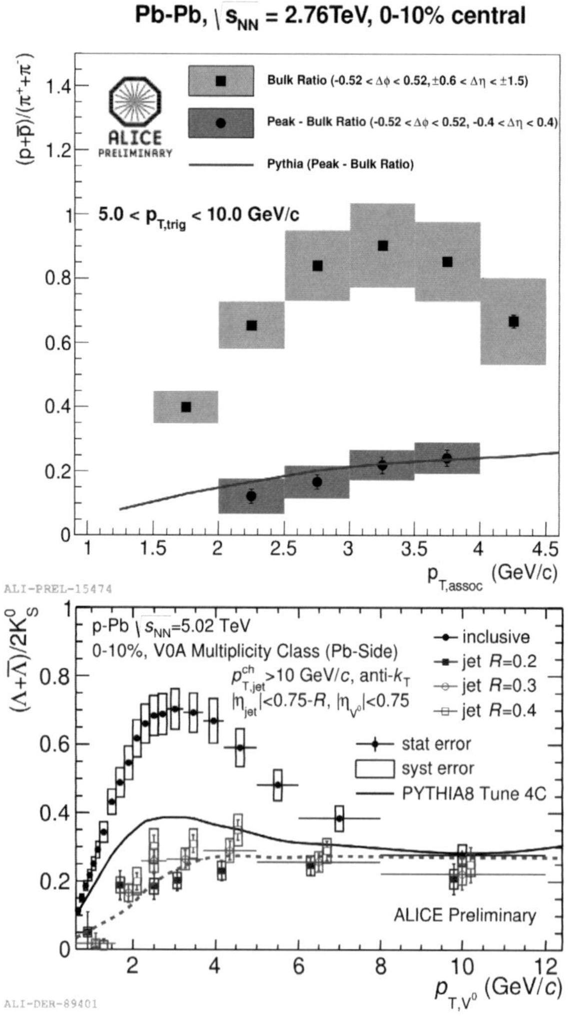

Figure 7. Top: p/π ratio as a function of associated particle p_T, in bulk and peak-bulk for 0-10% central Pb–Pb collisions at $\sqrt{s_{NN}} = 2.76$ TeV, with a leading (trigger) particle p_T between 5–10 GeV/c [11]. Bottom: Λ/K^0_S ratio in jet with different radii reconstructed with the anti-k_T method compared to PYTHIA8 and the Pb–Pb inclusive ratio (full black circles) for 0-10% central p–Pb collisions at $\sqrt{s_{NN}} = 5.02$ TeV.

Journal of Physics: Conference Series **636** (2015) 012009 doi:10.1088/1742-6596/636/1/012009

8. Conclusions

At high p_T we observe a suppression of identified particle production due to parton energy loss. The same suppression is seen for all light quark systems created in Pb–Pb collisions, which suggests that the chemical composition of leading particles from jets in the medium is similar to jets produced in vacuum. No suppression in p–Pb collisions is seen, indicating that the suppression observed in Pb–Pb is a final-state hot-matter effect.

At intermediate p_T, the particle ratios show a baryon-to-meson enhancement which in Pb–Pb is understood in the coalescence and/or hydrodynamic flow picture. In p–Pb collisions we see similar features, but less pronounced, as in Pb–Pb. By separating the underlying events from the jet-like structures, we note that the baryon-to-meson enhancement seems to be an effect arising in the underlying events in both Pb–Pb and p–Pb collisions, while the jet-like contributions appear to be unmodified.

References

[1] B. Abelev *at al.* (ALICE collaboration) 2013 *Phys. Rev. Lett.* **110** 082302
[2] B. Abelev *at al.* (ALICE collaboration) 2013 *Phys. Rev. Lett.* **111** 222301
[3] B. Abelev *at al.* (ALICE collaboration) 2014 *Phys. Lett.* B **736** 196-207
[4] P. Bozek 2012 *Phys. Rev.* C **85** 014911
[5] F. Barile (ALICE collaboration) 2014 *Nuclear Physics* A **926** 177-185
[6] T. Pierog *at al.* 2013 ArXiv ePrint: 1306.0121
[7] R. Fries *at al.* 2008 *Ann. Rev. Nucl. Part. Sci.* **58** 177-205
[8] B. Abelev *at al.* (ALICE collaboration) 2015 *Phys. Rev.* C **91** 024609
[9] B. Abelev *at al.* (ALICE collaboration) 2014 *Phys. Lett.* B **728** 025-038
[10] M. Veldhoen (ALICE collaboration) 2013 *Nuclear Physics* A **910-911** 306-309
[11] A. Ortiz Velasquez *at al.* 2013 *Phys. Rev. Lett.* **111** 042001

Journal of Physics: Conference Series **636** (2015) 012010 doi:10.1088/1742-6596/636/1/012010

Jet production in pp, p–Pb and Pb–Pb collisions measured by ALICE

Rosi Reed for the ALICE Collaboration

E-mail: rosi.reed@wayne.edu

Abstract. Particle jets, formed when a hard scattered parton fragments into a jet of hadrons, are an ideal probe of the medium formed in heavy-ion collisions. The hard-scattered partons that produce them come from early in the collision, prior to the medium formation. These partons lose energy as they traverse the medium, and eventually fragment into jets of hadrons, which exhibit a modification when compared to jets produced in pp and p–Pb collisions. At LHC energies, the parton production cross-section is much larger than at RHIC, allowing jets to be reconstructed over a much wider kinematic range. Jet reconstruction allows for a more differential investigation of the parton energy loss than single hadrons, which have been used as jet proxies in the past, as the jets collect a larger percentage of the final state energy, which means their kinematics are more closely correlated to the kinematics of the initial parton.

Jets are reconstructed in ALICE either using information from the tracking systems, or by combining this with the ALICE electromagnetic calorimeter (EMCal). In these proceedings, jet spectra from 2.76 TeV Pb–Pb and pp collisions will be presented. In particular, the centrality and event-plane dependence of the measured spectra and the background will be discussed. Jets from different centrality classes and event-plane orientations provide additional information necessary for understanding the path-length and temperature dependence of energy loss mechanisms. The reconstruction and correction procedures for jets will be shown. Results from Pb–Pb events will be compared to the baseline pp and p–Pb results, which allows the initial state and cold nuclear matter effects to be disentangled from hot medium effects. The jet nuclear modification, which quantifies the suppression, will be compared to energy-loss models.

1. Introduction

In ultra-relativistic heavy-ion collisions, a new state of matter is formed which is called the Quark Gluon Plasma (QGP), where quarks and gluons are deconfined. One way to characterize the properties of the QGP is to use partons, quarks or gluons, from the hard scattering of the partonic constituents in the colliding nucleons as probes of the medium. Hard scattering occurs early in the collision evolution, prior to QGP formation. The high transverse momentum (p_{T}) partons produced in these scatters propagate through the expanding medium, and eventually fragment into jets of hadrons. The high-p_{T} partons interact with the medium as they traverse through it, which will modify the fragmentation of the jets relative to proton-proton (pp) collisions. This modification is called jet quenching [1], and is the result of radiation or collisional energy loss of the high-p_{T} parton.

At the LHC, there are three collisional systems that are available for analysis: pp, p–Pb and Pb–Pb. Jet measurements in pp collisions are an important baseline for the heavy-ion collisions, because while the production cross section of the initial hard scattered partons is calculable using perturbative Quantum Chromodynamics (pQCD), the measurements allow the contribution from the non-perturbative hadronization processes to be calibrated. Measurements in p–Pb systems allow cold nuclear matter (CNM) effects to be quantified, which is key to understanding the modifications in Pb–Pb collisions.

2. Experimental Set-Up

The results reported in these proceedings are from data collected by the ALICE experiment in 2010 and 2011 at energies of 2.76 (pp and Pb–Pb), 5.02 (p–Pb), and 7 TeV (pp) per nucleon pair. Jets in ALICE are constructed from tracks measured in the ALICE tracking system and clusters measured by the Electromagnetic Calorimeter (EMCal). The tracks are from charged particles which are reconstructed with the Time Projection Chamber (TPC) and the Inner Tracking System (ITS), a six-layer silicon detector which provides precise measurement of the primary vertex location. The Silicon Pixel Detector (SPD), the two inner layers of the ITS, which has an acceptance in pseudorapidity of $|\eta_{\rm lab}| < 1.4$, is also used in order to select high quality tracks which correspond to tracks from the primary collision. Tracks are reconstructed at mid-rapidity ($|\eta_{\rm lab}| < 0.9$) and in full azimuth down to a transverse momentum ($p_{\rm T}$) of 0.15 GeV/c. The EMCal is a Pb-scintillator sampling calorimeter, which measures energy from neutral particles that interact electromagnetically, which consist mainly of photons and π^0s. The clusters are reconstructed at mid-rapidity $|\eta_{\rm lab}| < 0.7$ and in azimuth with an acceptance of $\Delta\varphi = 100^{\rm o}$ down to a cluster energy of 0.30 GeV. The ALICE VZERO scintillator detectors, which cover $2.8 < \eta_{\rm lab} < 5.1$ and $-3.7 < \eta_{\rm lab} < -1.7$, measure the forward particle multiplicity so they are used to determine the event centrality in Pb–Pb and p–Pb events. The VZERO detectors are also used to define the minimum-bias interaction trigger and also to select central events in Pb–Pb collisions. For a complete description of the ALICE detector see [2].

3. Jet reconstruction in ALICE

The jets measured by ALICE detector are placed into two categories: charged jets and full jets. The former are jets that are reconstructed only from charged particles measured with the ALICE tracking system, and are not corrected for the missing neutral energy. These jets are corrected for effects such as the tracking efficiency and resolution based on simulations of jets with PYTHIA [3, 4], with the detector response simulated by GEANT3 [5]. Full jets, which will be referred to as jets for the rest of these proceedings, include the neutral energy component measured by the EMCal, and these are corrected for all detector effects similarly as done for charged jets.

The collection of tracks and EMCal clusters corrected for charged particle contamination, are clustered into jets using the anti-$k_{\rm T}$ or the $k_{\rm T}$ algorithms [6, 7, 8, 9] with a resolution parameter of R = 0.2-0.6 and with the boosted $p_{\rm T}$ recombination scheme. In order to limit edge effects, only jets that are at least R away from the acceptance boundaries of the TPC or EMCal are used in the analyses described here. The jets found by the anti-$k_{\rm T}$ algorithm are used to determine the signal jets in the pp, p–Pb and Pb–Pb analyses, whereas the jets found by the $k_{\rm T}$ algorithm are used to quantify the underlying event density in the p–Pb and Pb–Pb analyses.

4. Results from pp

Jet measurements in pp collisions are needed as baseline measurements in order to quantify the modification of the jet spectrum in heavy-ion collisions due to the presence of the QGP. Two important baseline measurements are the differential jet cross-section in $\sqrt{s}$ = 2.76 TeV and the charged jet cross section at $\sqrt{s}$ = 7 TeV. The measured jets were corrected back to the particle level utilizing a simulation based bin-by-bin technique [10]. Figures 1 and 2 show the inclusive differential jet cross section at $\sqrt{s}$ = 2.76 TeV obtained with R = 0.2 and R = 0.4 compared to pQCD calculations at NLO with and without hadronization effects [11]. After the inclusion of hadronization and the parton shower, the calculations agree with the measured

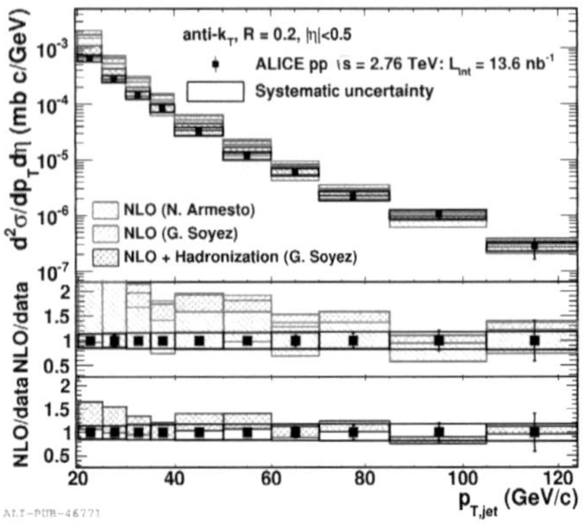

Figure 1. The inclusive differential jet cross section at $\sqrt{s} = 2.76$ TeV obtained with $R = 0.2$ compared to NLO pQCD calculations with and without hadronization effects [10].

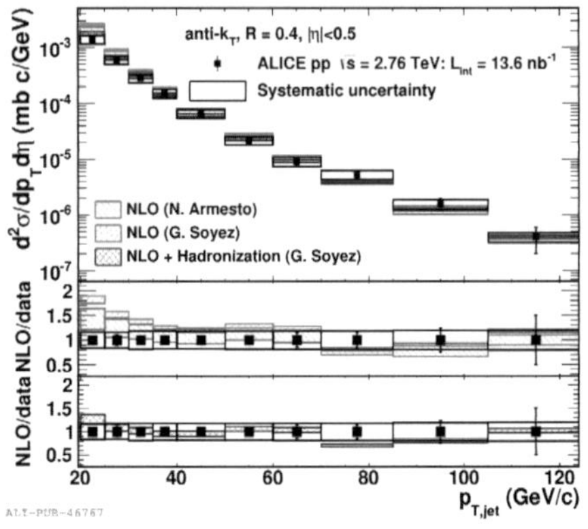

Figure 2. The inclusive differential jet cross section at $\sqrt{s} = 2.76$ TeV obtained with $R = 0.4$ compared to NLO pQCD calculations with and without hadronization effects [10].

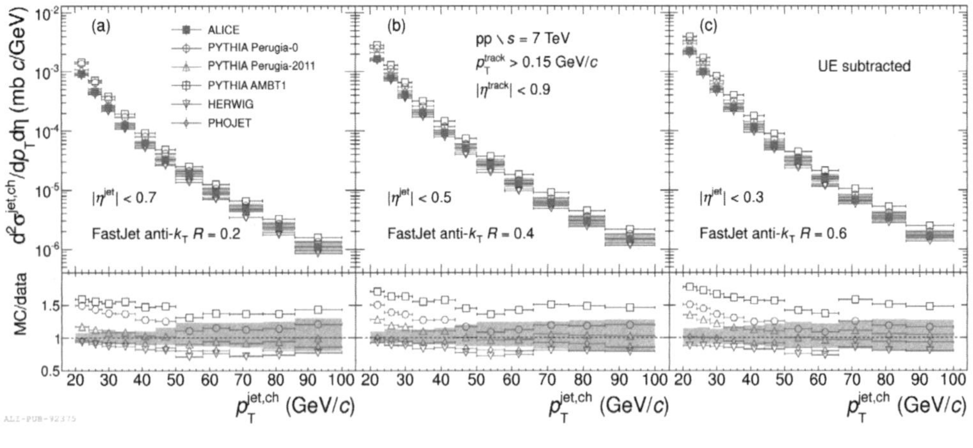

Figure 3. Charged jet cross sections measured in the ALICE experiment in pp collisions at $\sqrt{s} = 7$ TeV compared to several MC generators: PYTHIA AMBT1, PYTHIA Perugia-0 tune, PYTHIA Perugia-2011 tune, HERWIG, and PHOJET. Bottom panels: Ratios MC/Data. Shaded bands show quadratic sum of statistical and systematic uncertainties on the data drawn at unity [12].

spectrum, which indicates that jet formation is well understood. This is especially important for jets with a small cone size, which are often used in heavy-ion analyses as small jets are less affected by the underlying event than larger jets. Additional analysis details can be found in [10]. Figure 3 shows the inclusive differential charged jet cross section at $\sqrt{s} = 7$ TeV obtained with $R = 0.2$, 0.4 and 0.6 compared to PYTHIA and HERWIG simulations [12]. There is good agreement between the measured spectra and the Perugia 2011 tune of PYTHIA and HERWIG. This analysis is used to construct the pp reference for the 5.02 TeV jet R_{pPb}, as there is no

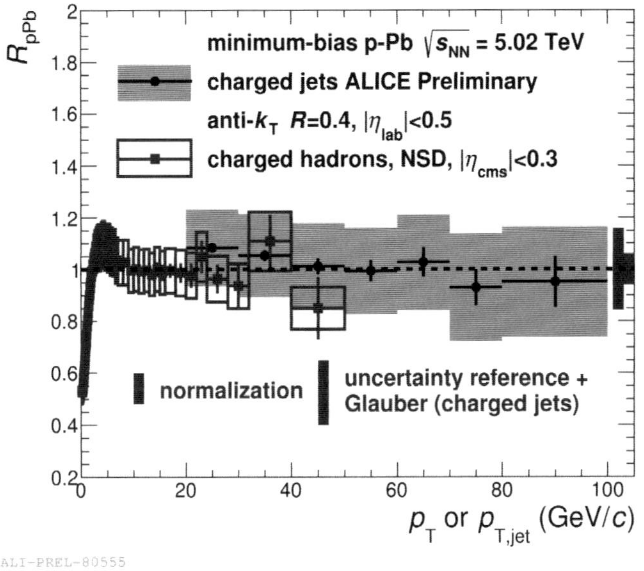

Figure 4. The solid circles show the $R_{\rm pPb}$ as a function of $p_{\rm T,\,jet}$ for R = 0.4 charged jets in minimum-bias p–Pb collisions at $\sqrt{s_{\rm NN}}$ = 5.02 TeV. The reference used was the measured spectrum from pp collisions at $\sqrt{s_{\rm NN}}$ = 7 TeV scaled using PYTHIA. The solid boxes around the points are the systematic uncertainties from the detector effects, background fluctuations and unfolding. Uncertainties in the reference and the Glauber model are shown as a small box about $R_{\rm pPb}$ =1 [13, 14]

pp data at 5.02 TeV.

5. Results from p–Pb

Cold nuclear matter (CNM) effects such as shadowing or the nuclear modification of the parton distribution function (PDF) can cause the jets measured in p–Pb to be modified compared to pp collisions. These measurements are necessary in order to quantify how much of the jet modification measured in Pb–Pb collisions is a result of hot nuclear matter effects. The results in these proceedings uses the data from p–Pb collisions at an energy of $\sqrt{s_{\rm NN}}$ = 5.02 TeV taken in the first months of 2013. The underlying event in p–Pb collisions is much less than in Pb–Pb collisions, however, a similar procedure was used to remove the contribution from the measured jet spectra. The background energy density, ρ, is calculated event-by-event and subtracted jet-by-jet using the formula $p_{T,jet}^{reco} = p_{T,jet}^{raw} - \rho \times A_{jet}$. For the full jet measurement, in order to reduce the effect of the limited EMCal acceptance, the charged track background density is measured in full azimuth and scaled using a factor that includes the electromagnetic contributions. The distribution of the region-to-region fluctuations around the mean background density is evaluated using the Random Cone (RC) approach. The distribution of $\delta p_{\rm T}$ is calculated by $\delta p_{\rm T} = \Sigma p_T - \rho \pi R^2$ for each event by randomly placing a cone with the same radius as the resolution parameter used for the jet finding algorithm into the acceptance and adding the transverse momenta of its constituents. The measured spectrum is corrected for background fluctuations and detector effects using an unfolding technique with a response matrix based on

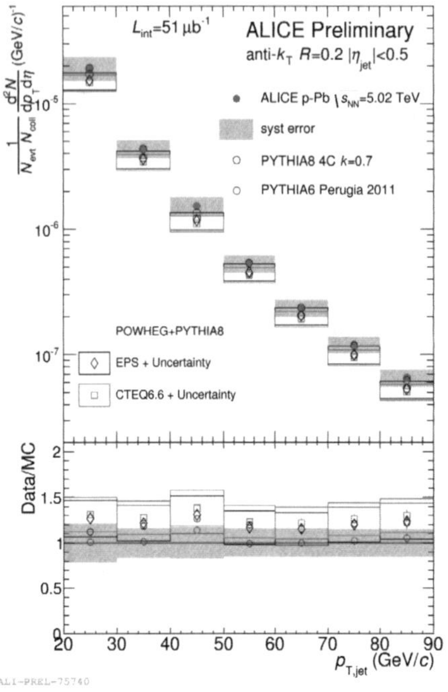

Figure 5. Fully corrected p-Pb jet spectrum for $R = 0.2$ scaled by $N_{\rm coll}$ compared to PYTHIA and POWHEG simulations at 5.02 TeV [15].

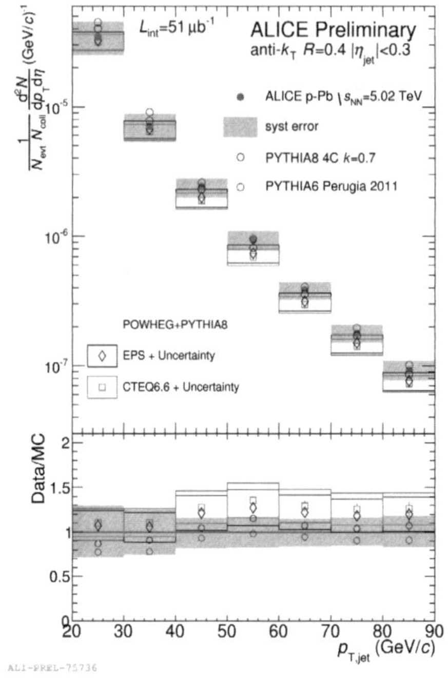

Figure 6. Fully corrected p-Pb jet spectrum for $R = 0.4$ scaled by $N_{\rm coll}$ compared to PYTHIA and POWHEG simulations at 5.02 TeV [15].

the measured $\delta p_{\rm T}$ distributions, and detector response from PYTHIA+GEANT simulations. The difference in the unfolded spectra between different algorithm choices, such as Bayesian and SVD, is taken as a systematic uncertainty.

The cold nuclear matter effects can be quantified by the nuclear modification factor, $R_{\rm pPb}$, from minimum bias collisions. This can be calculated as: $R_{\rm pPb}(p_{\rm T}) = \frac{dN_{pPb}/dp_T}{<T_{pPb}>d\sigma_{pp}/dp_T}$. The nuclear overlap function, $T_{\rm pPb}$, is calculated by $<T_{\rm pPb}> = <N_{\rm coll}> / \sigma_{\rm inelastic\ pp}$, where the number of binary collisions was calculated using a Glauber model. The reference for the charged jet $R_{\rm pPb}$ shown in Fig. 4 was created by scaling the measured $R = 0.4$ charged jet spectrum from 7 TeV shown in Fig. 3 using a bin-by-bin factor calculated with POWHEG NLO plus PYTHIA8 [13]. The scaling systematics were determined using PYTHIA, and a power-law interpolation-based scaling factor. For comparison, the charged hadron $R_{\rm pPb}$ is also shown [14]. The full jet spectra are shown for $R = 0.2$ and $R = 0.4$ jets in Figures 5 and 6. Calculations of PYTHIA, and NLO POWHEG with PYTHIA fragmentation agree well with the data [15]. The ratio of the data over the Monte Carlo, like the charged particle $R_{\rm pPb}$, is consistent with unity, which indicates that there is no large CNM effect on the jet spectrum in p–Pb collisions

6. Results from Pb–Pb

One of the main experimental challenges in heavy-ion collisions is removing the contribution from the underlying event from jet observables. As discussed in Section 5, this is done for the jet spectra by determining the average background energy density event-by-event and subtracting

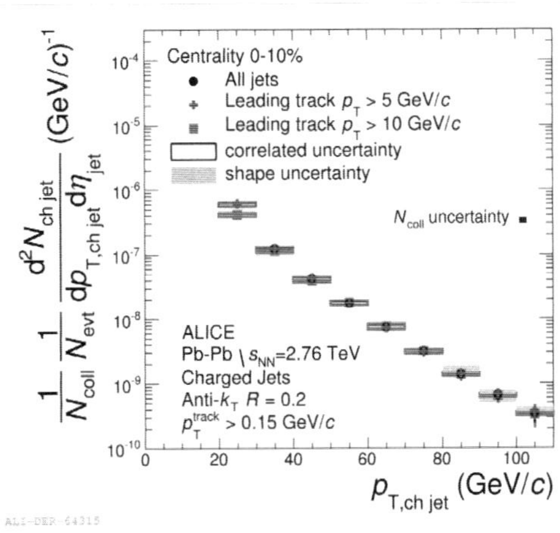

Figure 7. The inclusive charged jet spectra from the 10% most central events for $R = 0.2$, normalized by $N_{\rm coll}$ versus the choice of leading charged track $p_{\rm T}$ [16].

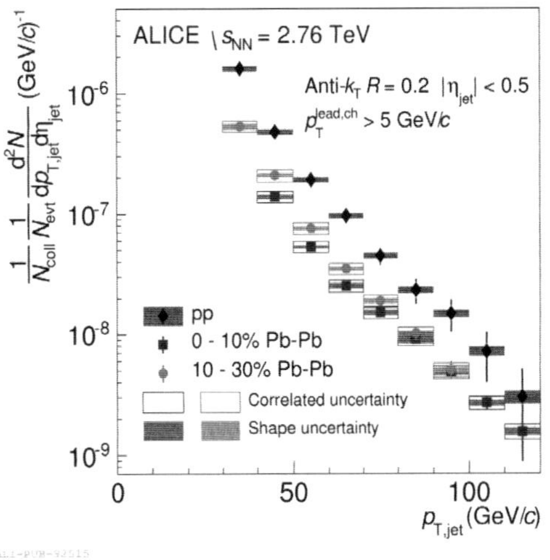

Figure 8. The inclusive $R = 0.2$ jet spectra with a 5 GeV/c leading charged track bias in 0-10%, 10-30%, and in pp collisions normalized by $N_{\rm coll}$[17].

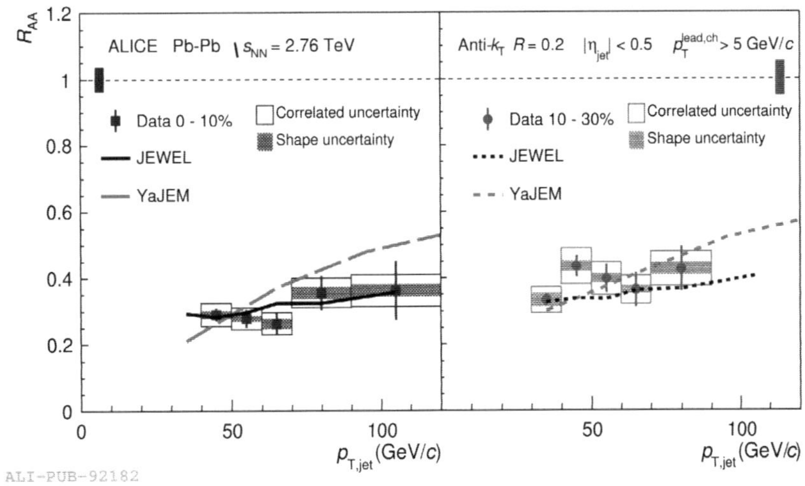

Figure 9. The $R = 0.2$ nuclear modification factor, $R_{\rm AA}$, for 0-10% and 10-30% central events compared to YaJEM and JEWEL [17]. The spectra used to determine the $R_{\rm AA}$ are shown in Figure 8.

it jet-by-jet. The geometric background fluctuations are also quantified by $\delta p_{\rm T}$, which is larger in central Pb–Pb collisions than it is in p–Pb and thus will smear the resulting $p_{\rm T,\ jet}$ spectrum more than for the same jet resolution parameter in p–Pb. The smearing of the jet $p_{\rm T}$ spectrum due to background fluctuations, as well as the smearing due to detector effects, is corrected on an ensemble basis in an unfolding procedure that is very similar to what was used in Section 5.

In order to remove combinatorial jets and achieve a stable result with the unfolding method for full jets, it was necessary to require that all jets have a track with $p_T > 5$ GeV/c. This selection does not introduce a large fragmentation bias above $p_{T,\text{ jet}} = 30$ GeV/c, and is effective at removing the combinatorial jets from the sample. The unfolded $R = 0.2$ charged jet spectra from the 10% most central events versus the choice of leading track bias is shown in Fig. 7. Above 30 GeV/c the effect is small. The $R = 0.2$ full jet spectra for 0-10% and 10-30% are shown in Fig. 8. Additionally, the biased pp spectrum is also shown, the analysis details are described in Section 4, other than the inclusion of the bias.

The nuclear modification factor, R_{AA}, for $R = 0.2$ full jets in 0-10% and 10-30% is shown in Fig. 9. Jets are suppressed with an R_{AA} of 0.28 ± 0.04 in 0–10% and 0.35 ± 0.04 in 10–30% collisions, independent of $p_{T,\text{ jet}}$ within both the statistical and systematic measurement uncertainties [17]. Also shown are the results from two different models, YaJEM [18] and JEWEL [19]. For a brief summary of the models, see [17]. Both calculations are found to reproduce the jet suppression measured in these events with calculated χ^2 of 1.690 for YaJEM and 0.368 for JEWEL, even though they take a very different approach to the inclusion of $\hat{q}$ into their calculations. In order to gain a better understanding of jet quenching, $R = 0.2$ charged jet spectra with a 3 GeV/c charged track bias relative to the reaction plane is measured, allowing for a determination of jet v_2. For this measurement, the average background and fluctuations are subtracted differentially with respect to the angle of the reaction plane, which decreases the p_T smearing of the resulting jets. The v_2 was corrected for the finite v_2 resolution using a three sub-event method [20]. The result in the 5% most central events can be seen in Fig. 10, where the expectation is that v_2 should be minimal given the geometry in these most central collisions. The result in 30-50% central events is shown in Fig. 11, where the path-length dependence should be maximal [20]. We do not see a significant indication of a non-zero jet v_2 in the most central events, but the semi-central events hint at a non-zero v_2, though are compatible with zero given the large uncertainties.

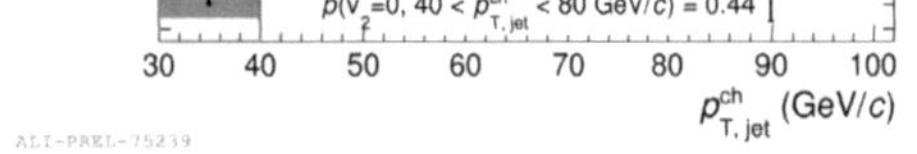

Figure 10. Charged jet v_2 in 5% most central Pb–Pb collisions at $\sqrt{s_{NN}} = 2.76$ TeV at mid-rapidity. The jets are required to have a leading charged hadron with p_T larger than 3 GeV/c [20].

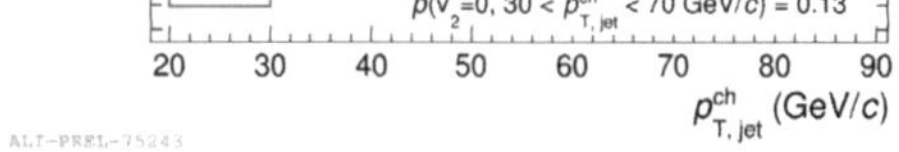

Figure 11. Charged jet v_2 in 30-50% central Pb–Pb collisions at $\sqrt{s_{NN}} = 2.76$ TeV at mid-rapidity. The jets are required to have a leading charged hadron with p_T larger than 3 GeV/c [20].

7. Conclusions

The jet spectra measurements from pp collisions at $\sqrt{s_{\rm NN}} = 2.76$ and 7 TeV by ALICE were shown. These results are important in establishing a baseline for cold nuclear and hot nuclear effects. In p–Pb collisions, we see no evidence of large cold nuclear matter effects at high $p_{\rm T}$, either compared to a pp reference scaled from $\sqrt{s}$ =7 TeV, or using PYTHIA and a NLO POWHEG and PYTHIA calculations. In Pb–Pb collisions jets are suppressed, with $R_{\rm AA} < 1$, with the largest suppression occurring in the most central events. Two models, YaJEM and JEWEL, give a fair comparison with the data. A jet v_2 measurement, which should allow a more differential understanding of jet quenching, is also shown, though drawing a significant conclusion is difficult with the current statistical and systematic precisions.

References

[1] U. A. Wiedemann, "Jet Quenching in Heavy Ion Collisions", Landolt-Bornstein **23**, 521 (2010) [arXiv:0908.2306 [hep-ph]].
[2] B. Abelev *et al.* [ALICE Collaboration], "Performance of the ALICE Experiment at the CERN LHC", Int. J. Mod. Phys. A **29**, 1430044 (2014) [arXiv:1402.4476 [nucl-ex]].
[3] T. Sjostrand, S. Mrenna and P. Z. Skands, "PYTHIA 6.4 Physics and Manual", JHEP **0605**, 026 (2006) [hep-ph/0603175].
[4] T. Sjostrand, S. Mrenna and P. Z. Skands, "A Brief Introduction to PYTHIA 8.1", Comput. Phys. Commun. **178**, 852 (2008) [arXiv:0710.3820 [hep-ph]].
[5] R. Brun, F. Bruyant, M. Maire, A. C. McPherson and P. Zanarini, "Geant3", CERN-DD-EE-84-1.
[6] M. Cacciari, G. P. Salam and G. Soyez, "The Anti-k(t) jet clustering algorithm", JHEP **0804**, 063 (2008) [arXiv:0802.1189 [hep-ph]].
[7] M. Cacciari and G. P. Salam, "Dispelling the N^3 myth for the k_t jet-finder", Phys. Lett. B **641**, 57 (2006) [hep-ph/0512210].
[8] M. Cacciari and G. P. Salam, "Pileup subtraction using jet areas", Phys. Lett. B **659**, 119 (2008) [arXiv:0707.1378 [hep-ph]].
[9] M. Cacciari, J. Rojo, G. P. Salam and G. Soyez, "Jet Reconstruction in Heavy Ion Collisions", Eur. Phys. J. C **71**, 1539 (2011) [arXiv:1010.1759 [hep-ph]].
[10] B. Abelev *et al.* [ALICE Collaboration], "Measurement of the inclusive differential jet cross section in pp collisions at $\sqrt{s} = 2.76$ TeV", Phys. Lett. B **722**, 262 (2013) [arXiv:1301.3475 [nucl-ex]].
[11] G. Soyez, "A Simple description of jet cross-section ratios", Phys. Lett. B **698**, 59 (2011) [arXiv:1101.2665 [hep-ph]].
[12] B. Abelev *et al.* [ALICE Collaboration], "Charged jet cross sections and properties in proton-proton collisions at $\sqrt{s} = 7$ TeV", arXiv:1411.4969 [nucl-ex].
[13] J. Adam *et al.* [ALICE Collaboration], "Measurement of charged jet production cross sections and nuclear modification in p–Pb collisions at $\sqrt{s_{\rm NN}} = 5.02$ TeV", arXiv:1503.00681 [nucl-ex].
[14] J. Adam *et al.* [ALICE Collaboration], "Centrality dependence of particle production in p–Pb collisions at $\sqrt{s_{\rm NN}} = 5.02$ TeV" arXiv:1412.6828 [nucl-ex].
[15] M. Connors [ALICE Collaboration], "Inclusive jet spectra in pPb collisions at ALICE", Nucl. Phys. A **931**, 1174 (2014) [arXiv:1409.3468 [nucl-ex]].
[16] B. Abelev *et al.* [ALICE Collaboration], "Measurement of charged jet suppression in Pb-Pb collisions at $\sqrt{s_{\rm NN}} = 2.76$ TeV", JHEP **1403**, 013 (2014) [arXiv:1311.0633 [nucl-ex]].
[17] J. Adam *et al.* [ALICE Collaboration], "Measurement of jet suppression in central Pb-Pb collisions at $\sqrt{s_{\rm NN}} =$ 2.76 TeV", Phys. Lett. B **746**, 1 (2015) [arXiv:1502.01689 [nucl-ex]].
[18] T. Renk, "Physics probed by the P_T dependence of the nuclear suppression factor", Phys. Rev. C **88**, no. 1, 014905 (2013) [arXiv:1302.3710 [hep-ph]].
[19] K. C. Zapp, "Geometrical aspects of jet quenching in JEWEL", Phys. Lett. B **735**, 157 (2014) [arXiv:1312.5536 [hep-ph]].
[20] S. Aiola [ALICE Collaboration], "Measurement of jet p_T spectra and R_{AA} in pp and PbPb collisions at $\sqrt{s_{NN}} = 2.76$ TeV with the ALICE detector", Nucl. Phys. A **931**, 382 (2014) [arXiv:1408.0479 [nucl-ex]].

Journal of Physics: Conference Series **636** (2015) 012011 doi:10.1088/1742-6596/636/1/012011

Measuring away-side jet modifications in Au+Au collisions at RHIC

Kun Jiang [12] **(for the STAR Collaboration)**

[1] Department of Modern Physics, University of Science and Technology of China, Hefei 230026, China

[2] Department of Physics and Astronomy, Purdue University, West Lafayette, IN, 47907, USA

E-mail: jiang304@purdue.edu

Abstract. We report measurements of jet correlations in Au+Au collisions at $\sqrt{s_{\rm NN}}$ = 200 GeV by the STAR experiment. In this analysis we devise a novel method to subtract flow background using data itself. The correlation width is studied as a function of centrality and associated particle p_T. The width is found to increase with centrality at modest to high associated particle p_T. The increase can arise from jet modification by medium and/or event averaging of away-side jets deflected by medium flow. The discrimination of the physics mechanisms requires further study by three-particle correlations.

1. Introduction

Jets are quenched in relativistic heavy ion collisions due to interactions in the dense medium [1, 2]. Jet-quenching can be measured via jet-like dihadron correlations. Measurements of jet medium modifications have so far been obscured because of the large underlying anisotropic flow background. The background subtraction was usually done by the ZYAM (zero yield at minimum) method [3] in the previous azimuthal di-hadron correlation measurements [2, 4]. Those measurements are usually coming with large uncertainties.

In this analysis, we attempt to subtract the flow background using the data itself. Fig. 1 shows a cartoon of the methodology. Di-hadron correlations are analyzed in two η regions symmetric about midrapidity, one (close-region) close to and the other (far-region) far away from the P_x (Eq. 1) cut η window. The away-side jet contributes to the close-region but not as much to the far-region due to the large $\Delta\eta$ gap. The correlation difference measures the away-side jet shape where the anisotropic flow background is cleanly subtracted.

2. Event and track selections

This analysis uses 4.3×10^8 minimum-bias-triggered (MB) Au+Au events at the nucleon-nucleon center-of-mass energy of $\sqrt{s_{\rm NN}} = 200$ GeV taken by the STAR experiment in 2011. The minimum bias trigger condition is defined as a coincidence signal between the east and west vertex position detectors (VPD) [5] located at $4.4 < |\eta| < 4.9$. The main detector used for this analysis is the Time Projection Chamber (TPC) [6]. Particle tracks are reconstructed in the TPC. The events are required to have the reconstructed primary vertex within 30 cm of the TPC center along the beam direction. To remove secondary tracks from particle decays, only tracks that extrapolate to within 2 cm of the primary vertex are used. Tracks are required to be reconstructed with

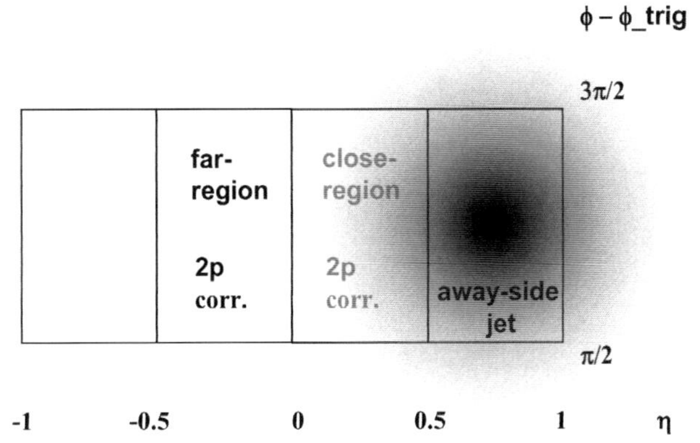

Figure 1. Cartoon of the methodology. We select events with a large recoil P_x (Eq. 1) from a high-p_T trigger particle within a given η window (in the cartoon, $0.5 < \eta < 1$) to enhance the away-side jet population. Di-hadron correlations are analyzed in two η ranges symmetric about midrapidity, one ($0 < \eta < 0.5$) close to and the other ($-0.5 < \eta < 0$) far away from the P_x cut η window. Anisotropic flow background are the same in close-region and far-region.

at least 20 out of a maximum of 45 hits in the TPC. The ratio of the number of hits used in track reconstruction to the number of possible hits is required to be greater than 0.51 to eliminate multiple track segments being reconstructed from a single particle trajectory. The track pseudo-rapidity cut is $|\eta| < 1$ in the TPC.

3. Data analysis

We select events with a large recoil P_x from a high-p_T trigger particle within a given η window to enhance the away-side jet population. P_x is defined as

$$P_x|_{\eta_1 \leq \eta \leq \eta_2} = \sum_{\cos(\phi - \phi_{trig})<0} p_T \cdot \cos(\phi - \phi_{trig})/\epsilon. \quad (1)$$

Only away-side particles are included in the P_x calculation so that the η distribution of the trigger particle is unbiased by this selection. Single-particle acceptance and detection efficiency ϵ is applied. The ϕ-dependent correction is obtained from the inverse of the single-particle ϕ distribution whose average is normalized to unity. This correction has been done as a function of time and data with similar efficiency are grouped together. The η-dependent correction is obtained by treating symmetrized $dN/d\eta$ distributions from z_{vtex} range $|z_{vtex}| < 2$ cm as the baseline. The ratio of the $dN/d\eta$ distribution from each z_{vtex} bin to this baseline would be the η- and also z_{vtex}-dependent corrections.

Fig. 2 shows P_x distributions from a high-$p_T$trigger particle with $3 < p_T^{trig} < 10$ GeV/c. The distributions from central and peripheral collisions are different because of the number of tracks. For each centrality, we cut on the left tail of the distribution (10% of events) to enhance the away-side jet population.

The azimuthal dihadron correlation is given by

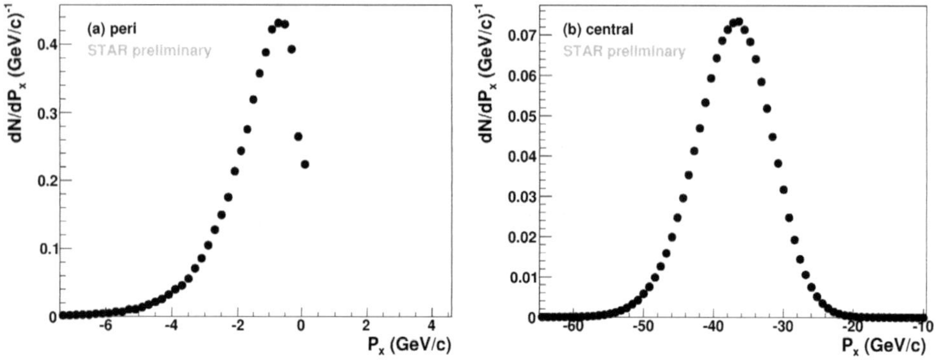

Figure 2. Distributions of recoil P_x from a high-p_T trigger particle with $3 < p_T^{trig} < 10$ GeV/c for peripheral (70% − 80%) and central (0% − 5%) collisions.

$$\frac{1}{N_{trig}}\frac{dN}{d\Delta\phi} = \frac{1}{N_{trig}}\int d\Delta\eta \frac{S(\Delta\eta,\Delta\phi)}{B(\Delta\eta,\Delta\phi)}. \tag{2}$$

where $S(\Delta\eta,\Delta\phi) = \frac{1}{N_{trig}}\frac{d^2N}{d\Delta\eta d\Delta\phi}$ and $B(\Delta\eta,\Delta\phi) = \frac{d^2N}{d\Delta\eta d\Delta\phi}$ are the signal and mixed-event background distributions. The mixed-event background serves as the correction for the detector two-particle acceptance. The mixed-event background distributions are normalized to have an average of unity where the two-particle $\Delta\eta$ acceptance is 100%. The ϕ-averaged single particle efficiency is not corrected because this analysis deals with only the correlation shape, not the absolute amplitude.

4. Results

The left plot in Fig. 3 shows the dihadron azimuthal correlations for close-region and far-region in min-bias Au+Au collisions at $\sqrt{s_{\rm NN}} = 200$ GeV. The near-side peaks are almost identical while

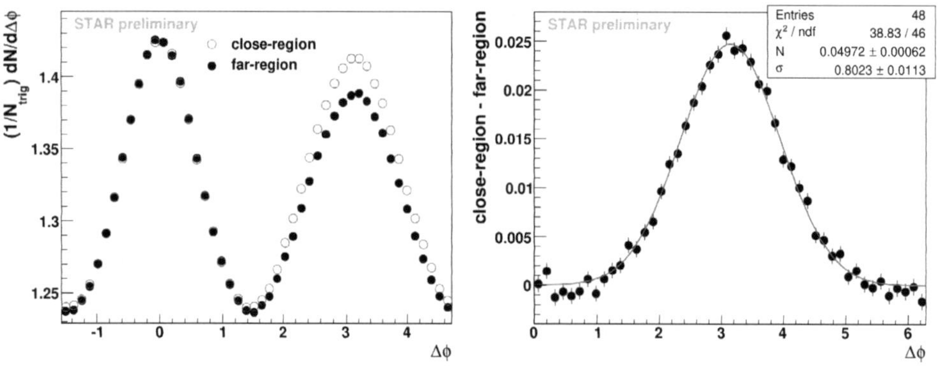

Figure 3. (left) Dihadron azimuthal correlations in close-region (open circles) and far-region (solid circles) for $3 < p_T^{trig} < 10$ GeV/c and $1 < p_T^{assoc} < 2$ GeV/c in min-bias Au+Au collisions at $\sqrt{s_{\rm NN}} = 200$ GeV. (right) The difference between close- and far-regions azimuthal correlations. The x-axes of the histograms are adjusted to show the full shape of the away-side peak ($\Delta\phi = \pi$). The curve is a Gaussian fit to the away-side jet shape with the mean value fixed at π.

the away-side peaks are different. The difference is from away-side jet contributions. The right plot in Fig. 3 shows the difference between close-region correlations and far-region correlations, which measures the away-side jet correlation shape. A Gaussian fit is applied to extract the away-side jet correlation width. A good χ^2 per degree of freedom (ndf) is obtained.

Fig. 4 shows away-side jet correlation width as a function of centrality for various p_T^{assoc}. The systematic uncertainties are estimated by varying P_x cut from allowing 10% of events to 2%, 5%, 15%, 20%, 30% and 50% of events. The P_x cut gives a systematic error of 3.1%. An additional 3.8% tracking systematic uncertainty is obtained by varying track quality cuts.

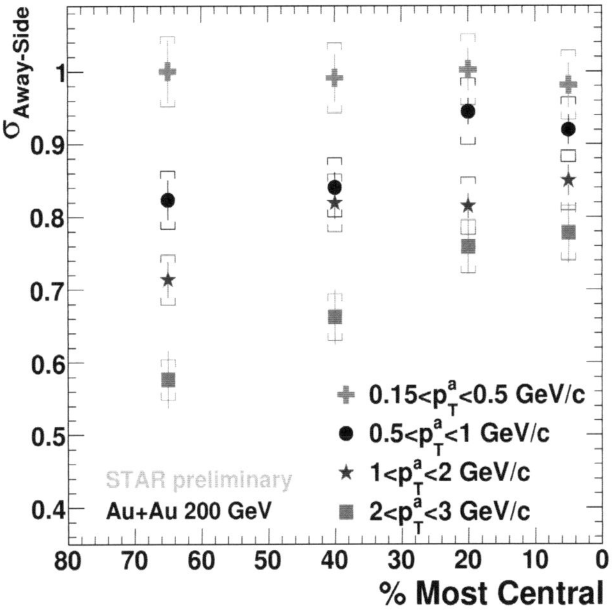

Figure 4. Away-side jet correlation width as a function of centrality for $3 < p_T^{trig} < 10$ GeV/c and various p_T^{assoc} bins in Au+Au collisions at $\sqrt{s_{NN}} = 200$ GeV. The horizontal caps indicate the systematic errors.

5. Conclusions

A novel method was devised to measure away-side jet correlations with a clean, robust flow subtraction using data itself. The away-side jet correlation width broadens with increasing centrality at modest to high p_T^{assoc} and is insensitive to centrality at low p_T^{assoc}. The possible physics mechanisms are, among others, jet modification and jet deflection. Discrimination of these mechanisms requires further study, such as 3-particle correlations [7].

References

[1] Adams J *et al.* (STAR) 2003 *Phys.Rev.Lett.* **91** 072304 (*Preprint* nucl-ex/0306024)
[2] Adams J *et al.* (STAR) 2005 *Phys.Rev.Lett.* **95** 152301 (*Preprint* nucl-ex/0501016)
[3] Ajitanand N, Alexander J, Chung P, Holzmann W, Issah M *et al.* 2005 *Phys.Rev.* **C72** 011902 (*Preprint* nucl-ex/0501025)
[4] Aggarwal M *et al.* (STAR) 2010 *Phys.Rev.* **C82** 024912 (*Preprint* 1004.2377)
[5] Llope W, Geurts F, Mitchell J, Liu Z, Adams N *et al.* 2004 *Nucl.Instrum.Meth.* **A522** 252–273 (*Preprint* nucl-ex/0308022)

[6] Anderson M, Berkovitz J, Betts W, Bossingham R, Bieser F *et al.* 2003 *Nucl.Instrum.Meth.* **A499** 659–678 (*Preprint* nucl-ex/0301015)
[7] Abelev B *et al.* (STAR) 2009 *Phys.Rev.Lett.* **102** 052302 (*Preprint* 0805.0622)

Journal of Physics: Conference Series **636** (2015) 012012 doi:10.1088/1742-6596/636/1/012012

Jet quenching studies in CMS

Hallie Causey Trauger for the CMS Collaboration

Department of Physics, University of Illinois at Chicago
845 W. Taylor St. M/C 273, Chicago, IL 60607

E-mail: hallie.causey.trauger@cern.ch

Abstract. CMS has a wide set of results exploring the radiation patterns of jets in the quark gluon plasma and the medium response to jets, through studies of correlations between charged hadrons and jets. This talk reviews the various CMS results that examine the phenomena from different perspectives, and reports on the recent additions to these results. In particular, a study of dijet momentum imbalance, studied differentially in distance from dijet axis, sets limits on the propagation of the radiated momentum, and a study of the jet-track correlation functions examines the extent to which the radiated energy remains correlated with the jet axes.

1. Introduction

The modification of highly energetic probes traversing the quark gluon plasma, known as "jet quenching," has been firmly established at RHIC and LHC [1, 2]. High-precision measurements at the LHC have furthermore shown that the detailed structures within the jet cone (radius parameter $R = 0.3$) are modified by the medium in terms of both p_{T} and angular distributions [3, 4, 5, 6]. However, these observed in-cone changes only explain a small fraction of the centrality-dependent dijet momentum imbalance, indicating that a large amount of energy is radiated outside of the jet cone or to very low momentum particles. Extending such measurements to large angles is crucial to properly constrain the energy loss mechanisms and to quantify the interaction strength within the QGP. Taking advantage of the kinematic reach of hard probes and the availability of detailed characterization of the event bulk properties, the CMS detector is able to separate the perturbative quantum chromodynamics (QCD) components of jet-track correlations from the hydrodynamic background to explore the two-dimensional energy flow about the jet axis beyond the limit of the small cone radius. Here, we present two complementary analyses that extend previous measurements to large angles in relative pseudorapidity ($\Delta\eta$) and relative azimuth ($\Delta\phi$), while providing a more detailed look at energy redistribution about high transverse momentum jets.

First, presenting results from CMS analysis [7], we analyze overall momentum imbalance between the leading and subleading sides of high-p_{T} dijets. This technique measures subleading-to-leading excesses in track multiplicity and transverse momentum when comparing the total energy in the hemispheres associated with the subleading versus leading jet. Next, presenting new results from CMS analysis [8], we study charged particle per-trigger yields correlated to high-p_{T} jets, looking for energy redistribution both inside and outside of the jet cone. This allows for analysis of modifications and energy flow on each side of the dijet individually, as well as for an inclusive sample of high-p_{T} jets.

Journal of Physics: Conference Series **636** (2015) 012012 doi:10.1088/1742-6596/636/1/012012

2. Data Samples and Jet and Track Reconstruction

The analyses presented here use data collected with the CMS detector (described in detail in [9]) at a center-of-mass energy of 2.76 TeV, with PbPb collisions taken in 2011 with an integrated luminosity of $166\mu b^{-1}$, and pp reference taken in 2013 with an integrated luminosity of 5.3 pb^{-1}. The events for these analyses were selected using the CMS High Level Trigger (HLT), an inclusive single-jet trigger that selects $p_T > 80$ GeV/c. The HLT is fully efficient for events containing reconstructed jets with $p_T > 120$ GeV/c satisfying offline selection as documented in [2, 10]. The dijet samples in both analyses are selected by the following critera: events are first required to contain a leading (highest p_T) jet in the pseudorapidity range of $|\eta| < 2$ with a corrected jet $p_T > 120$ GeV/c and a subleading (next-highest p_T) jet of $p_T > 50$ GeV/c, also in $|\eta| < 2$. The azimuthal angle between the leading and subleading jets is required to be at least $5\pi/6$. No explicit requirement is made either on the presence or absence of a third jet in the event. To ensure stable jet reconstruction performance and good tracker acceptance for tracks on all sides of each jet, only events in which both leading and subleading jets fall within $|\eta| < 1.6$ are included in the final selected data sample. For the dijet imbalance study, jets are further restricted to $|\eta| < 0.5$ so that tracks within $|\Delta\eta| < 1.9$ of a jet fall within the tracker coverage of $|\eta| < 2.4$; the correlation study instead corrects for geometrical pair-acceptance effects.

For both pp and PbPb collisions, jet reconstruction is performed with the anti-k_T algorithm with distance parameter $R = 0.3$, that is encoded in the FastJet framework [11, 12]. Jets are reconstructed based on energy deposits in the CMS calorimeters. For PbPb collisions the CMS algorithm "HF/Voronoi" is used to predict and subtract the heavy ion underlying event based on information from HF energy measurements as well as Voronoi decomposition of particle flow [13, 7]. For pp collisions no underlying event subtraction is employed. For track reconstruction in PbPb data, we employ an iterative charged particle reconstruction capable of reconstructing tracks with transverse momentum as low as 0.4 GeV/c. For studies of pp data, charged particles are reconstructed using the same iterative method [14] as in the previous CMS analyses of pp collisions. Detailed descriptions of track reconstruction and studies of tracking efficiency corrections can be found in Ref. [7].

Monte Carlo (MC) event generators have been used for evaluation of the jet and track reconstruction performance, to establish the jet energy scale (JES), and to correct for tracking efficiency. Jet events are generated by the PYTHIA MC generator [15] (version 6.423, tune Z2 [16]). These generated PYTHIA events are propagated through the CMS detector using the GEANT 4 package [17] to simulate the detector response. In order to account for the influence of the underlying PbPb events, the PYTHIA events are embedded into fully simulated PbPb events, generated by HYDJET [18] (version 1.8) that is tuned to reproduce the total particle multiplicities, charged hadron spectra, and elliptic flow at all centralities. Detailed information about sources of systematic uncertainty in the dijet imbalance and jet-track correlation analyses is provided in [7] and [8] respectively.

3. Results: Dijet Momentum Imbalance

Detailed information about the overall momentum balance in dijet events can be obtained using the projection of the p_T of reconstructed charged particles onto an axis in the azimuthal plane (ϕ_{Dijet}) midway between the direction of the leading jet and the opposite direction of the subleading jet. As the leading and subleading jets are required to be approximately back-to-back, this axis is very close to that for the leading jet ($\phi_{\mathrm{Dijet}} \approx \phi_1$). For each event, this projection was calculated as

$$\not{p}_{\mathrm{T}}^{\parallel} = \sum_{\mathrm{i}} -p_{\mathrm{T}}^{\mathrm{i}} \cos(\phi_{\mathrm{i}} - \phi_{\mathrm{Dijet}}), \qquad (1)$$

where the sum is over all tracks with $p_T > 0.5$GeV/c and $|\eta| < 2.4$. The results were then averaged over events in a given A_J bin to obtain $\langle \not{p}_T^{\parallel} \rangle$.

Figure 1 shows $\langle \not{p}_T^{\parallel} \rangle$ for pp collisions and four selections of PbPb collision centrality. For each panel, $\langle \not{p}_T^{\parallel} \rangle$ is plotted in four bins of dijet asymmetry A_J ranging from almost balanced to very unbalanced pairs. On each panel in the upper row, $\langle \not{p}_T^{\parallel} \rangle$ averaged over all tracks with $p_T > 0.5$GeV/c, and the contributions to A_J from individual ranges of track p_T are shown. Following the definition of $\langle \not{p}_T^{\parallel} \rangle$, negative values indicate an excess in the direction of the leading jet, while positive values indicate an excess in the direction of the subleading jet. When selecting events containing dijets with $A_J > 0.11$, an excess of high p_T particles in the direction of the leading jet (indicated by the red boxes) is seen for all selections in pp and PbPb collisions, as expected. For pp and peripheral PbPb collisions, this excess is mostly balanced by intermediate p_T particles from $2-8$ GeV/c. Going to more central collisions, the balancing contribution shifts from the intermediate p_T range towards low p_T particles in the $0.5-2$ GeV/c range. This effect is most pronounced for large A_J dijets seen in central PbPb collisions, where pathlength effects are expected to be largest. In the lower row of Fig. 1, the difference between the contributions of different track p_T ranges to $\langle \not{p}_T^{\parallel} \rangle$ in PbPb and pp collisions is shown after requirements on the PbPb collision centrality and the dijet imbalance. While the contributions from the various p_T ranges are similar for to pp reference peripheral collisions, a difference can be seen for central collisions, with a significant excess of low p_T charged particles for asymmetric jets in PbPb collisions.

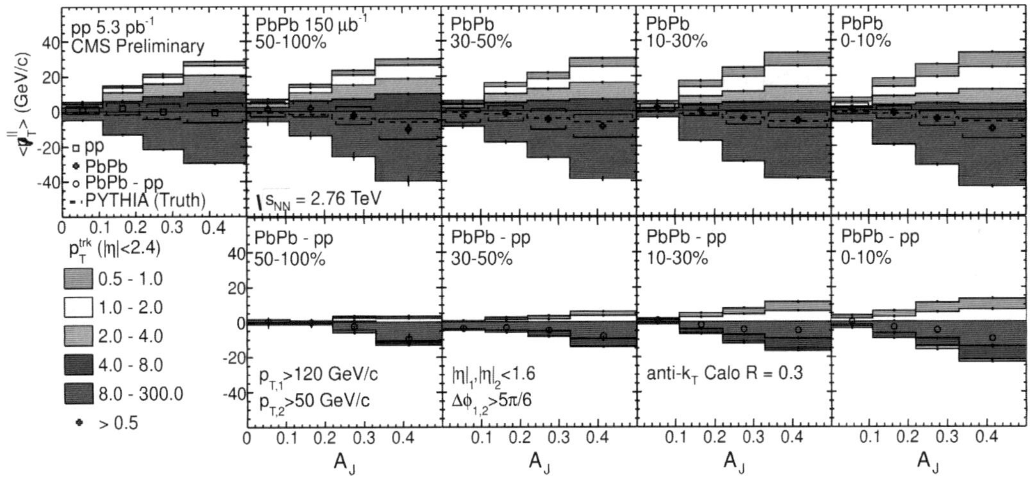

Figure 1. Upper row shows average missing transverse momentum, $\langle \not{p}_T^{\parallel} \rangle$, for pp collisions (left) and four selections of PbPb collision centrality. The solid markers show $\langle \not{p}_T^{\parallel} \rangle$ averaged over all tracks with $p_T > 0.5$GeV/c, while the colored boxes show the contribution to $\langle \not{p}_T^{\parallel} \rangle$ for various momentum ranges from $0.5 < p_T < 1$GeV/c (light blue) to $p_T > 8$GeV/c (red). For each panel $\langle \not{p}_T^{\parallel} \rangle$ values are shown as a function of dijet asymmetry from almost balanced ($A_J < 0.11$) to unbalanced ($A_J > 0.33$) dijets. Vertical bars and brackets represent the statistical and systematic uncertainties, respectively. The lower row shows the difference PbPb−pp of the $\langle \not{p}_T^{\parallel} \rangle$ contribution for the individual momentum ranges shown in the upper panel. Error bars and brackets represent statistical and systematic uncertainties respectively.

A key question for understanding the physics of QCD radiation in the parton energy loss process concerns the angular distribution of the energy loss relative to the axis defined by the parton direction. To study this question, the contribution to the overall $\langle \not{p}_{\mathrm{T}}^{\parallel} \rangle$ from annular regions around the leading and subleading jet axes is extracted. Annular regions in $\Delta R = \sqrt{\Delta\phi^2_{\mathrm{Trk,jet}} + \Delta\eta^2_{\mathrm{Trk,jet}}}$ of width 0.2 from $0 < \Delta R < 0.2$ to $1.6 < \Delta R < 1.8$ are defined, with an additional region for $1.8 < \Delta R < 2.9$. For each region, $\langle \not{p}_{\mathrm{T}}^{\parallel} \rangle$ for tracks falling into the region is calculated in the momentum ranges defined previously. For large ΔR, the annuli around the leading and subleading jet may overlap. In this case, each track in the overlap region is only counted once in the $\langle \not{p}_{\mathrm{T}}^{\parallel} \rangle$ sum, with the track belonging to the annulus of smaller radius.

The analysis is performed for pp collisions and for two PbPb centrality selections, $30 - 100\%$ and $0 - 30\%$. The resulting differential $\langle \not{p}_{\mathrm{T}}^{\parallel} \rangle$ spectra as a function of ΔR are shown in the upper row of Fig. 2, for pp and PbPb collisions. The plots show a large $\langle \not{p}_{\mathrm{T}}^{\parallel} \rangle$ contribution from high-p_{T} particles towards the leading jet for small radii $\Delta R < 0.2$, with balancing contributions from low-p_{T} particles towards the subleading jet for larger ΔR. The cumulative $\langle \not{p}_{\mathrm{T}}^{\parallel} \rangle$ values (i.e., integrating the total $\langle \not{p}_{\mathrm{T}}^{\parallel} \rangle$ over ΔR starting at $\Delta R = 0$) are shown as dashed lines for pp and solid lines for PbPb, demonstrating the evolution of the overall momentum balance from small to large distances to the dijet axis. A comparison of pp and PbPb collisions is provided in the lower row of Fig. 2, showing the difference in the contributions of various momentum ranges to $\langle \not{p}_{\mathrm{T}}^{\parallel} \rangle$ and the difference of the total $\langle \not{p}_{\mathrm{T}}^{\parallel} \rangle$ vs. ΔR. For peripheral events, the momentum spectrum of the $\langle \not{p}_{\mathrm{T}}^{\parallel} \rangle$ contributions is nearly identical between pp and PbPb. For central PbPb events, however, the PbPb-pp difference shows an excess of low momentum tracks ($0.5 - 2$GeV/c) towards the subleading jet side for all regions with $\Delta R > 0.2 - 0.4$. While the momentum distribution in each ΔR bin is different between pp and central PbPb, the overall angular dependence of the balancing contribution is similar, as is seen in the nearly ΔR independent PbPb-pp difference of the total $\langle \not{p}_{\mathrm{T}}^{\parallel} \rangle$ plotted as the open circles.

4. Results: Jet-Track Correlations

To further study modifications to the particle yields associated with with individual jets (leading and subleading sides of the dijet as well as an inclusive sample of high-p_{T} jets), we turn to a complementary analysis of correlations between charged hadrons and reconstructed jets. In this study, all charged tracks in the event with p_{T} above 1 GeV/c are used to construct a two-dimensional relative pseudorapidity ($\Delta\eta = \eta_{\mathrm{jet}} - \eta_{\mathrm{track}}$), relative-azimuth ($\Delta\phi = \phi_{\mathrm{jet}} - \phi_{\mathrm{track}}$) correlation with respect to the measured jet axis for inclusive jets and for leading and subleading jets in unbalanced dijets. The jet-track correlations are constructed following the procedure established in [19], for the following bins in charge particle transverse momentum: $1 < p_{\mathrm{T}}^{\mathrm{assoc}} < 2$ GeV/c, $2 < p_{\mathrm{T}}^{\mathrm{assoc}} < 3$ GeV/c, $3 < p_{\mathrm{T}}^{\mathrm{assoc}} < 4$ GeV/c, and $4 < p_{\mathrm{T}}^{\mathrm{assoc}} < 8$ GeV/c. To correct for detector acceptance effects, raw correlation plots are divided by correlations between jets and minimum bias events (following the method of previous works [20, 21, 22]). The long-range contributions of the full 2D correlation is estimated by the $\Delta\phi$ projection of this correlation over the range $1.5 < |\Delta\eta| < 3.0$. The fit to this $\Delta\phi$ distribution is propagated uniformly in $\Delta\eta$, and subtracted from the acceptance-corrected yield. This background-subtracted yield is projected into $\Delta\eta$ and $\Delta\phi$ respectively. Mixed event correction and background subtraction are described in detail in [8].

The projections of the 2D correlations for leading and subleading jets onto $\Delta\eta$ (projected over $|\Delta\phi| < 1.0$) are shown in Fig. 3, for the lowest charged particle momentum selection. The upper panel of each figure shows the centrality evolution of the correlations for inclusive jets with p_{T} >120 GeV/c. In every panel a reference measurement from pp data at the same collision energy is shown with open symbols. To better visualize the PbPb to pp comparisons,

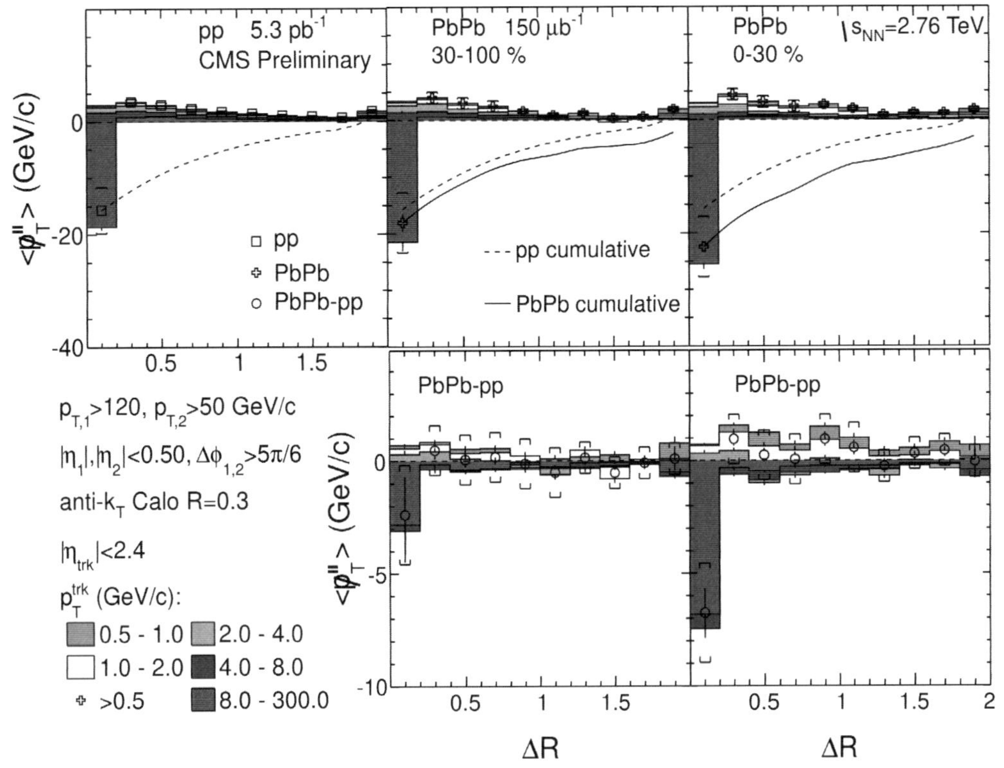

Figure 2. Upper row shows differential missing p_T distributions for pp, 30-100% and 0-30% PbPb data for various p_T ranges (colored boxes), as a function of ΔR. Also shown is the total $\langle \not{p}_T^{\parallel} \rangle$ (i.e., integrated over all p_T for a given ΔR bin) as a function of ΔR for pp (open squares) and PbPb data ($\times$ symbols). Dashed lines (pp) and solid lines (PbPb) show the cumulative $\langle \not{p}_T^{\parallel} \rangle$. The lower row shows the difference between the PbPb and pp differential $\langle \not{p}_T^{\parallel} \rangle$ distributions per p_T range as a function of ΔR (colored boxes) and difference of the total $\langle \not{p}_T^{\parallel} \rangle$ as a function of ΔR (open circles). Differences of cumulative $\langle \not{p}_T^{\parallel} \rangle$ are not shown. Error bars and brackets represent statistical and systematic uncertainties respectively.

the difference of the PbPb and pp correlation distributions is presented in the bottom panel for all centralities. Correlations are symmetrized in $\Delta\eta$ and display. Projections of leading and subleading jets onto $\Delta\phi$ (projected over $|\Delta\eta| < 1.0$) may be found in reference [8], and are very similar in width and shape to the $\Delta\eta$ correlations shown.

For the most peripheral events studied (centrality 50-100%), the PbPb correlations at low transverse momentum $1 < p_T^{\rm assoc} < 2$ GeV$/c$ remain similar to those of pp reference events for our jet selections, with perhaps a hint of a small additional yield in PbPb collisions. This hint turns into a stronger signal as a function of collision centrality, with the most significant excess present in the most central collisions. The observed excess in the low-p_T per-trigger particle yields extends to large angles far beyond the jet cone parameter $\Delta R = 0.3$, and exhibits a Gaussian-like shape relative to the jet direction in both $\Delta\eta$ and $\Delta\phi$. We note that these observations are consistent with the CMS results of jet-shape modifications (Ref. [5]) and fragmentation function measurements (Ref. [6]), and furthermore extend both of these measurements outside of the jet-cone radius parameter of 0.3 enabling a more detailed assessment of the medium-induced modifications.

For this lowest track p_T bin we observe that the excess of correlated yield extends significantly beyond the typical jet reconstruction radius. The excess of the soft correlated yields is more pronounced on the more "quenched" subleading side. Detailed cross-checks have been performed that confirm the consistency of these results with missing transverse momentum measurements discussed above. To summarize our findings and quantify the total yield excess observed in the PbPb data with respect to the pp reference at the same energy, we plot the integrals of the PbPb-pp yield differences as a function of track p_T and collision centrality in Fig. 4. As this figure shows, in both leading and subleading jets the excess yield diminishes for higher momentum tracks until the yield becomes similar to or even less than that of the pp reference for the highest track p_T bin of 4–8 GeV/c.

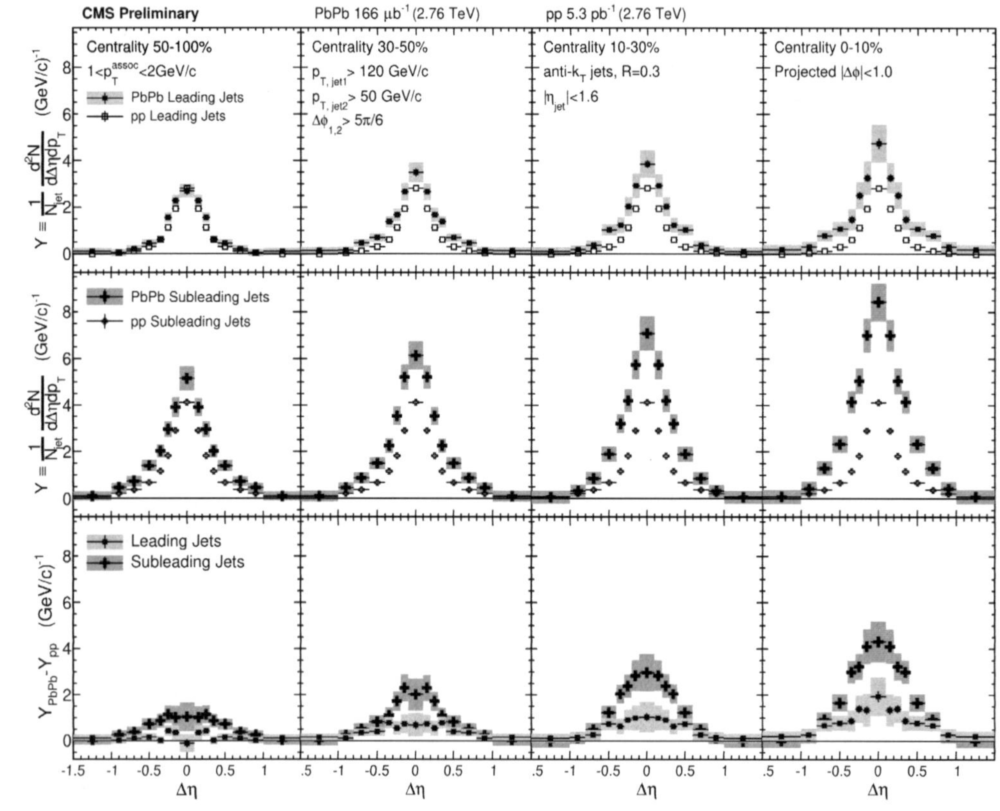

Figure 3. The top panel shows the $\Delta\eta$ distributions of charged particle background-subtracted yields correlated to PbPb and pp leading jets with $p_T > 120$ GeV/c. The middle panel shows the same distributions for subleading jets, and the bottom panel shows the difference PbPb minus pp for both leading and subleading jets. The total systematic uncertainties are shown as shaded boxes.

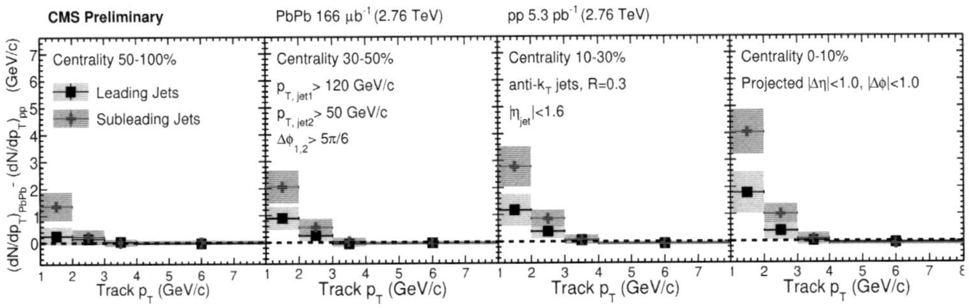

Figure 4. Total yield excess observed in the PbPb data with respect to the reference measured in pp collisions is shown as a function of track p_{T} in four different centrality intervals (0–10%, 10–30%, 30–50%, 50–100%) for both leading and subleading jets. The total systematic uncertainties are shown as shaded boxes.

To characterize the widths in $\Delta\eta$ and $\Delta\phi$ of the charged particle distributions, we fit the measured correlations with a double Gaussian function (found to best describe the overall correlation shapes). The width is defined as the $\Delta\eta$ or $\Delta\phi$ value that contains 67% of the total correlated yield. Width uncertainties are calculated by determining widths of $\Delta\eta$ and $\Delta\phi$ distributions varied by their respective systematic uncertainties, which are treated as fully correlated for the purposes of this determination. Widths are obtained for $\Delta\eta$ and $\Delta\phi$ projections of correlations to leading and subleading jets in CMS analysis [8]. Here for demonstration widths are shown as a function of track p_{T} for $\Delta\eta$ correlations to leading jets only, in Fig. 5.

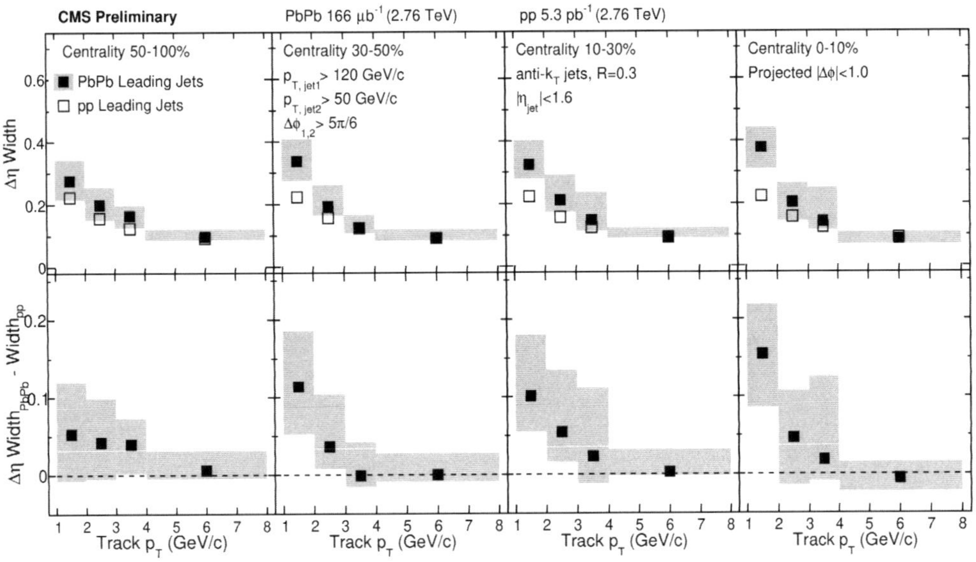

Figure 5. Width of the $\Delta\eta$ distributions after background subtraction as a function of track p_{T} comparing PbPb to pp data for leading jets with $p_{\mathrm{T}} > 120$ GeV/c. The shaded band corresponds to the systematic uncertainty.

5. Summary

Energy redistribution about high transverse momentum jets in heavy ion collisions has been studied using two complementary techniques. First, in a study of dijet momentum imbalance between leading and subleading hemispheres of an event, the missing-p_T balancing momentum distribution was studied as a function of dijet asymmetry and radial distance from dijet axis. In these results the subleading-to-leading energy balance showed a centrality-dependent redistribution of transverse momentum from high-p_T particles within the jet cone into low-p_T particles extending to large angles in radius parameter ΔR. Next, in an analysis of two-dimensional $\Delta\eta$–$\Delta\phi$ correlations, a centrality-dependent excess of correlated yield in PbPb collisions with respect to pp reference was observed for both leading and subleading jets, as well as for an inclusive jet sample. This excess yield (PbPb - pp) grows with centrality and decrease with transverse momentum, and remains correlated with jet axis while extending to large angles for all jet samples studied. Together, these two sets of results provide a comprehensive evaluation of medium modifications to jet properties, extending measurements to large angles in relative pseudorapidity ($\Delta\eta$) and relative azimuth ($\Delta\phi$) far beyond the typical jet cone radii.

References

[1] Aad G *et al.* (ATLAS) 2010 *Phys. Rev. Lett.* **105** 252303 (*Preprint* 1011.6182)
[2] Chatrchyan S *et al.* (CMS) 2012 *Phys.Lett.* **B712** 176–197 (*Preprint* 1202.5022)
[3] Chatrchyan S *et al.* (CMS) 2011 *Phys. Rev. C* **84** 024906 (*Preprint* 1102.1957)
[4] Chatrchyan S *et al.* (CMS) 2012 *JHEP* **1210** 087 (*Preprint* 1205.5872)
[5] Chatrchyan S *et al.* (CMS) 2014 *Phys. Lett. B.* **730** 243–263
[6] Chatrchyan S *et al.* (CMS) 2014 *Phys.Rev.* **C90** 024908 (*Preprint* 1406.0932)
[7] 2014 *CMS Physics Analysis Summary* **CMS-PAS-HIN-14-010**
[8] 2015 *CMS Physics Analysis Summary* **CMS-PAS-HIN-14-016**
[9] 2008 *JINST* **3** S08004
[10] Chatrchyan S *et al.* (CMS) 2013 *Phys.Lett.* **B718** 773–794 (*Preprint* 1205.0206)
[11] Cacciari M, Salam G P and Soyez G 2011 (*Preprint* 1111.6097)
[12] Cacciari M, Salam G P and Soyez G 2008 *JHEP* **04** 063 (*Preprint* 0802.1189)
[13] 2013 *CMS Detector Performance Summary* **CMS-DP-2013-018**
[14] 2010 *CMS Physics Analysis Summary* **CMS-PAS-TRK-10-001**
[15] Sjöstrand T, Mrenna S and Skands P 2006 *JHEP* **05** 026 (*Preprint* hep-ph/0603175)
[16] R Field 2010 (*Preprint* 1010.3558)
[17] Agostinelli S *et al.* 2003 *Nucl. Inst. Meth.* **A506** 250
[18] Lokhtin I P and Snigirev A M 2006 *Eur. Phys. J. C* **45** 211 (*Preprint* hep-ph/0506189)
[19] Collaboration C (CMS) 2013 *Phys. Lett. B* **724** 213 – 240
[20] Collaboration C (CMS) 2010 *JHEP* **09** 091
[21] Chatrchyan S *et al.* (CMS) 2011 *JHEP* **1107** 076 (*Preprint* 1105.2438)
[22] Collaboration C (CMS) 2013 *Phys. Lett. B* **718** 795–814

31st Winter Workshop on Nuclear Dynamics (WWND2015) IOP Publishing
Journal of Physics: Conference Series **636** (2015) 012013 doi:10.1088/1742-6596/636/1/012013

Decay of Resonaces in Strong Magnetic Field

Peter Filip
Institute of Physics, Slovak Academy of Sciences, Dúbravská cesta 9, Bratislava 82103, Slovakia
E-mail: peter.filip@savba.sk

Abstract. We suggest that decay properties (branching ratios) of hadronic resonances may become modified in strong external magnetic field. The behavior of $K^{\pm *}$, K^{0*} vector mesons as well as $\Lambda^*(1520)$ and Ξ^{0*} baryonic states is considered in static fields 10^{13}- 10^{15} T. In particular, $n = 0$ Landau level energy increase of charged particles in the external magnetic field, and the interaction of hadron magnetic moments with the field is taken into account. We suggest that enhanced yield of dileptons and photons from $\rho^0(770)$ mesons may occur if strong decay channel $\rho^0 \to \pi^+\pi^-$ is significantly suppressed. CP - violating $\pi^+\pi^-$ decays of pseudoscalar η_c and $\eta(547)$ mesons in the magnetic field are discussed, and superpositions of quarkonium states $\eta_{c,b}$ and $\chi_{c,b}(nP)$ with $\Psi(nS)$, $\Upsilon(nS)$ mesons in the external field are considered.

1. Introduction

Modification of $\rho(770)$ decay properties in the magnetic field has been considered in relation to the possibility of $\rho^{\pm}$ meson condensation [1] in sufficiently strong static magnetic fields. The paper [1] suggests that the lowest energy of charged decay products ($\pi^+\pi^-$) is expected to exceed the mass of ρ^0 meson at field strength $eB \approx 0.14$ GeV^2. This means $\rho^0 \to \pi^+\pi^-$ decays become suppressed and thus ρ^0 meson can be "stabilized" in the magnetic field background [1].

One may immediately consider the possibility that strong decays of other hadrons can also be influenced in the magnetic field. Modification of decay probability into certain decay channels gives different branching ratios (BR) in the magnetic field when compared to the vacuum case. This may potentially affect reconstructed yields of resonances in heavy ion collisions, where very strong magnetic fields are expected to be formed for a very short time [2], and reconstructed yields of the observed hadronic resonance can thus become decay-channel-dependent.

In Sections 3, 4, 5 and 6 of this contribution we investigate to what extent $\rho^0(770)$, $K^*(892)$, $\Lambda^*(1520)$ and $\Xi^{0*}(1535)$ resonances may be affected by magnetic fields of strength 10^{13}- 10^{15} T, which may be present in collisions of heavy nuclei [2] at RHIC and LHC.

Because decay $\rho^0 \to \pi^+\pi^-$ is the only possible strong decay channel of ρ^0 meson, the enhancement of dilepton and photon production from ρ^0 decays can occur if $\rho^0 \to \pi^+\pi^-$ decay is closed for whatever reason. In Section 4 we speculate that excess of photons and e^+e^- or $\mu^+\mu^-$ pairs from ρ^0 decays may be expected in the magnetic field of sufficient strength.

Decay properties of $(q\bar{q})$ bound states in the magnetic field may be influenced also via different mechanism, which involves quantum superposition of $J = 0$ singlet and $L = 1$ states with $J = 1$ ($m_z = 0$ and $m_z = \pm 1$) triplet states observed in Positronium [3, 4]. If such phenomenon happens in the case of hadrons [5], it may lead also to the enhancement of CP - violating decays $\eta \to \pi^+\pi^-$ in the magnetic field. This interesting behavior is discussed in Sections 7 and 8.

2. Energy of charged particles in strong magnetic field

In the magnetic field, charged particle energy becomes quantized and for the lowest possible $n = 0$ Landau level [6] the energy of electrons or positrons can be expressed as

$$E[B, s_z] = \sqrt{m^2 + p_z^2 + |Q|B - 2QBs_z} + \Delta E[B, s_z] \quad (1)$$

where m, Q are mass and charge of the particle, p_z is momentum parallel to $\vec{B}$ field, and ΔE is a correction for the anomalous part of magnetic moment (see below). For $p_z^2 \ll m^2$, $Q = -e$ and $eB \ll m^2$ (assuming here $\Delta E \approx 0$) one obtains (e is electron charge magnitude)

$$E[B, s_z] \simeq m + p_z^2/2m + eB/2m + eBs_z/m\,. \quad (2)$$

Term eBs_z/m on the right side of Eq.(2) corresponds to the energy of magnetic dipole in external field $E = -\vec{\mu} \cdot \vec{B}$ with Dirac moment value $\mu = s_z Q/m$, and term $eB/2m$ accounts for the lowest Landau level [6]. In Figure 1 we show how the energy of static ($p_z = 0$) charged pions and kaons increases in magnetic field due to $n = 0$ Landau level energy according to Eq.(1). Dashed lines are obtained using Equation (2) and dotted lines show the behavior of neutral ($J = 0$) mesons.

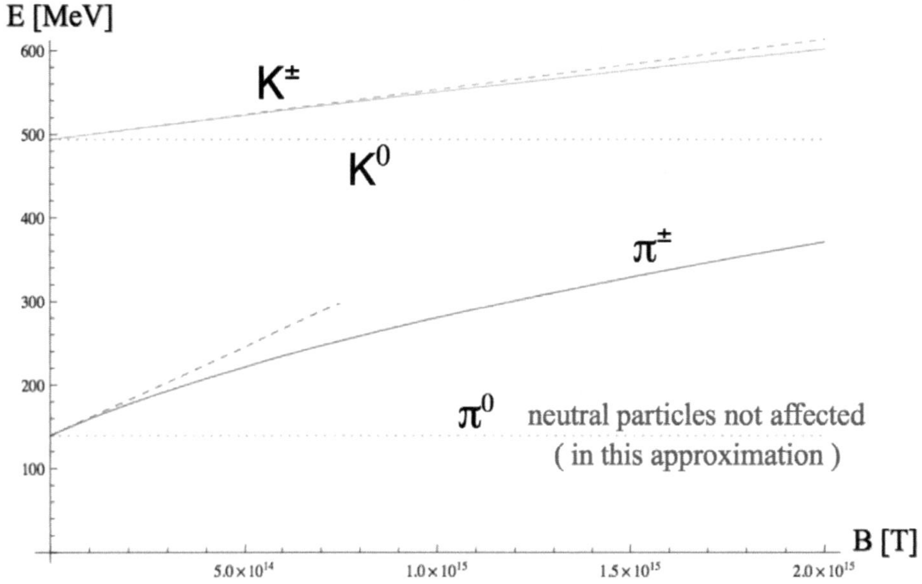

Figure 1. Energy of $\pi^\pm$ and $K^\pm$ mesons (with $p_z = 0$) in the magnetic field.

One can also read from Eq.(1) that energy of electrons with $s_z = -1/2$ and positrons with $s_z = +1/2$ does not change with magnetic field (if one neglects anomalous $\approx 10^{-3}$ part of their magnetic moments).

Particles with internal structure have gyromagnetic ratio g substantially different from Dirac value ($g = 2$) and "anomalous" magnetic moment needs to be taken into account properly. Energy dispersion relation for neutron in the magnetic field [7] is

$$E[B, s_z] = \left[p_z^2 + \left(\sqrt{m_n^2 + p_T^2} - (2s_z)\kappa_n \mu_N B\right)^2\right]^{1/2} \quad (3)$$

where $\kappa_n = -1.91$ is neutron magnetic moment in units $\mu_N = e\hbar/2m_p = 3.15 \cdot 10^{-14}$ MeV/T. For small magnetic fields: $E[B, s_z] \approx m_n + (2s_z)|\kappa_n|\mu_N B$ is obtained from Eq.(3) using $(p_z, p_T \approx 0)$, which corresponds to static magnetic dipole interaction $E = -\vec{\mu}_n \cdot \vec{B}$. We shall use Eq.(3) also for Λ and Ξ^0 baryons with their magnetic moments $\mu_{\Lambda^0} = -0.61\mu_N$ and $\mu_{\Xi^0} = -1.25\mu_N$. For neutral $J = 3/2$ particles $\Lambda^*(1520)$ and $\Xi^{0*}(1530)$, term $(2s_z)$ in Eq.(3) can replaced by $(2s_z/3)$ for obtaining the energy of $s_z = \pm 3/2$ substates.

Dispersion relation for protons in the magnetic field [7] can be expressed for $n = 0$ Landau level as

$$E[B, s_z] = \left[p_z^2 + m_p^2\left(\sqrt{1 + (1 - 2s_z)B/B_c^p} - s_z\kappa_p B/B_c^p\right)^2\right]^{1/2} \tag{4}$$

where B_c^p constant is $m_p^2c^2/e\hbar = 1.48 \cdot 10^{16}$ T and $\kappa_p = 1.79$ is anomalous part of proton magnetic moment. In weak magnetic fields ($B \ll B_c^p$) one gets $E[B, s_z] \approx m_p + eB/2m_p - (2s_z)[1 + \kappa_p]\mu_N B$ for $p_z \approx 0$, which corresponds well also to Eq.(2) using $\mu_p = (1 + \kappa_p)\mu_N$.

Behavior of electrons in very strong magnetic fields was calculated by J. Schwinger, and for $B > 10^{11}$ T the correction term ΔE in Eq.(1) is [7]

$$\Delta E[B, s_z] = 2s_z m_e(\alpha/4\pi)[\ln(2B/B_c^e) - (C + 3/2)]^2 + \ldots \tag{5}$$

where $\alpha \simeq 1/137$, $B_c^e = m_e^2c^2/e\hbar = 4.414 \cdot 10^9$ T, and $C = 0.577$ is Euler's constant. For fields $B < B_c^e$, the correction for anomalous magnetic moment is $\Delta E = 2s_z(\alpha/4\pi)m_e B/B_c^e$, which corresponds to $\Delta E \approx 2s_z 10^{-3}|\mu_e|B$. For muon one can replace $m_e \to m_\mu$ in the equations, which gives $B_c^\mu = B_c^e(207)^2 = 1.89 \cdot 10^{14}$ T, and for $B \ll B_c^\mu$ we have $\Delta E \approx 2s_z 10^{-3}|\mu_\mu|B$.

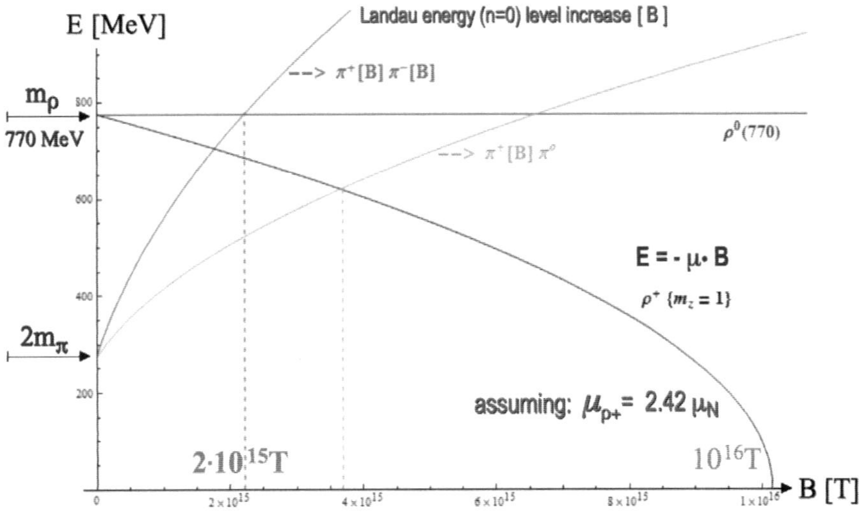

Figure 2. Energy of ρ^0 and ρ^+ ($s_z = +1$) mesons [1] and $\pi^+\pi^{-(0)}$ pairs in magnetic field.

3. Strong decays of $\rho(770)$ mesons in the magnetic field

Let us summarize the mechanism [1] which allows the lifetime of $\rho(770)$ mesons to be prolonged due to the suppression of $\rho \to \pi\pi$ strong decays. Energy of charged pseudoscalar ($s = 0$) mesons increases in magnetic field according to Eq.(1) and therefore at some critical field value B^{cr} the mass of decay products may become equal to the mass of decaying neutral hadron. For $\rho^0(770)$ meson this occurs in field $B^{cr} \approx 2.2 \cdot 10^{15}$ T, as shown in Figure 2. Decay $\rho^0 \to \pi^+\pi^-$ becomes therefore energetically forbidden in fields $B > B^{cr}$.

Magnetic moment μ_{ρ^0} of ρ^0 meson $(u\bar{u} - d\bar{d})/\sqrt{2}$ state is usually [1] considered to be zero, neglecting in this way quadratic Zeeman interaction [5] of $\rho^0(s_z{=}0)$ substate (see the behavior of ortho-Positronium (e^+e^-) $J = 1$ state in magnetic field [8]). However, charged $\rho^+ = (u\bar{d})$ and $\rho^- = (d\bar{u})$ mesons should have anomalous magnetic moments, which can be estimated also from the simple constituent quark model. For $u, \bar{d}$ quarks with parallel spins in $\rho^+(J = 1)$ meson ground state, one has $\mu_{\rho^+} = (\mu_u + \mu_{\bar{d}}) = 1.85\mu_N + 0.97\mu_N = \pm 2.82\mu_N$.

In Figure 2 we show energy decrease of $s_z = +1$ spin projection of charged ρ^+ mesons evaluated according to publication [1], where condensation of $\rho^\pm$ mesons is suggested to occur in static magnetic field. For majority of our estimates, however, it will be sufficient to consider weak-field approximations of the exact energy dispersion relations presented in Section 2.

4. Enhanced γ and l^+l^- production from ρ^0 meson decays

In the case of $\rho^0 \to \pi^+\pi^-$ decay suppression, the only potentially feasible strong decay channel $\rho^0 \to \pi^0\pi^0$ is forbidden by C parity and isospin conservation. Therefore, closing of $\pi^+\pi^-$ decay channel ($BR = 99\%$ in vacuum) has intriguing consequences: other decay channels of type $\rho^0 \to \pi^0\gamma$ ($BR = 6 \cdot 10^{-4}$) and $\rho^0 \to \eta\gamma$ ($BR = 3 \cdot 10^{-4}$) may become strongly enhanced as shown in Table 1. This means decays of ρ^0 meson in sufficiently strong magnetic field will produce excessive photons (from channels containing π^0, η) and possibly also dilepton pairs.

Table 1. ρ^0 decay branching ratios (BR) in B field.

channel	BR in $B \approx 0$	BR in $B > 2{\cdot}10^{15}$ T
$\pi^+\pi^-$	99%	0
$\pi^+\pi^-\gamma$	0.9%	0
$\pi^0\pi^0$	0	0*
$\pi^0\gamma$	$6 \cdot 10^{-4}$	64%
$\eta\gamma$	$3 \cdot 10^{-4}$	31%
l^+l^-	$0.9 \cdot 10^{-4}$	$\lesssim$1%
$\pi^0\pi^0\gamma$	$0.4 \cdot 10^{-4}$	4%
$\pi^+\pi^-\pi^0$	$1 \cdot 10^{-4}$	0
$\pi^+\pi^-\pi^+\pi^-$	$0.2 \cdot 10^{-4}$	0

*Assuming $\rho^0 \to \pi^0\pi^0$ remains closed in the magnetic field.

Regarding dilepton pair production via $\rho^0 \to l^+l^-$ decays in the strong magnetic field, one can make the following simplified consideration: $J = 1$ angular momentum of decaying ρ^0 meson must be conserved during decay process and this is achieved by parallel spin orientation (e.g. $\uparrow\uparrow$ or $\downarrow\downarrow$ for $J_z = \pm 1$) of produced leptons l^+l^-. From Eq.(1) it is then clear that energy $E[B, s_z]$ of one of the leptons will be increasing with the magnetic field, while energy of the other lepton remains almost constant (corrections ΔE from Eq.(5) are negligible). At field $B = 2.5{\cdot}10^{15}$ T the Equation (1) gives $E[B, s_z = +1/2] \approx 550$ MeV for electrons or muons, which allows $\rho^0 \to l^+l^-$ decays to happen with restricted phase-space (energy of l^+l^- pair does not exceed ρ^0 mass).

Directions of the momenta of produced charged leptons leaving the region of overcritical field B^{cr} will be affected by the Lorentz force $F = q \cdot \vec{v} \times \vec{B}$, which distorts (influences) the reconstructed invariant mass of dilepton pairs. Therefore $dN/dM_{l^+l^-}$ distribution of dileptons from the considered ρ^0 meson decays (in volume containing strong magnetic field $B \approx B^{cr}$) should/can be different from the expected $dN/dM_{l^+l^-}$ spectrum in ($B = 0$) vacuum case.

It is interesting to point out, that excessive elliptic flow asymmetry v_2 of photons has been observed by PHENIX collaboration [9] in Au+Au collisions. A simulation with enhanced decay probability of $\rho^0 \to \gamma + \pi^0$ and $\rho^0 \to \gamma + \eta$ channels could easily clarify, whether enlarged v_2 of photons observed [9] may be explained by suppressed (for whatever reason) $\rho^0 \to \pi^+\pi^-$ decays.

Assuming, that e.g. only 3% of ρ^0 mesons produced in heavy ion collision have $\pi^+\pi^-$ decay channel closed, branching ratios from the right side of Table 1 can easily generate the excess of photons or dileptons many times above the expected yield, without noticeably lowering the amount of $\rho^0 \to \pi^+\pi^-$ decays from the remaining (97% unaffected) ρ^0 mesons.

5. The case of K^{0*} and $K^{\pm*}$ meson decays

Mesons $K^{0*}(d\bar{s})$ and $\bar{K}^{0*}(\bar{d}s)$ are $(J = 1)$ particles of mass $M = 896$ MeV, which decay via strong decay channels $\bar{K}^{0*}(K^{0*}) \to K^{\pm} + \pi^{\mp}$ and $K^{0*} \to K^0 + \pi^0$ with probabilities 66% and 33% (ratio 2:1 is determined by Clebsch-Gordan coefficients originating from the isospin conservation). In Figure 3 we show the energy of $(K^0\pi^0)$ and $(K^{\pm}\pi^{\mp})$ decay products and of decaying K^{0*} meson in the magnetic field. Triplet splitting observed for K^{0*} state comes from the magnetic moment of K^{0*} meson, which can be estimated to be $\mu_{K^{0*}} = (\mu_d + \mu_{\bar{s}}) = -0.36\mu_N$ assuming parallel spin orientation for $d, \bar{s}$ quarks in $1S$ quantum state. Interaction energy of (K^{0*}) particle with magnetic field is evaluated using expression $E[B, J_z] = m_{K^{0*}} - BJ_z\mu_{K^{0*}}$ which agrees with Eq.(3) in small field limit, when $(2s_z) \to J_z$ substitution is used and $p = 0$.

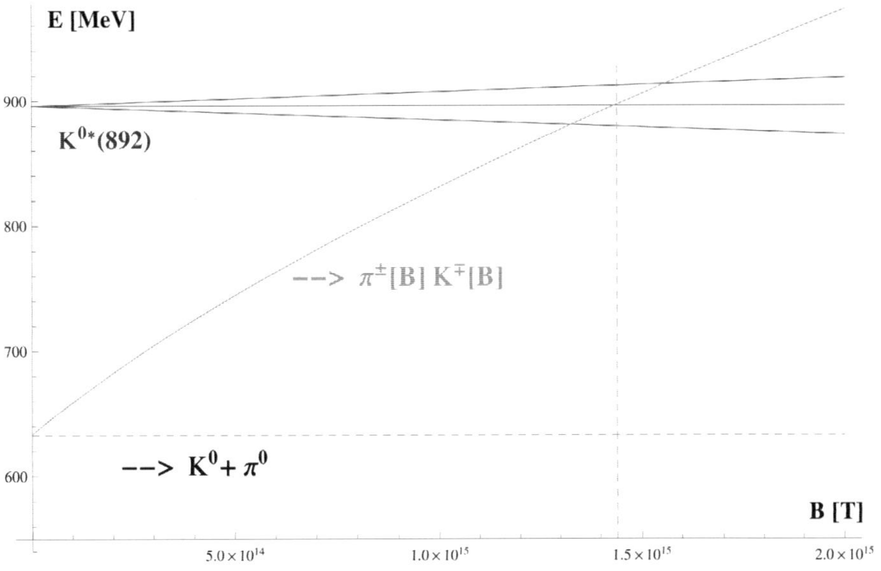

Figure 3. Energy of $K^{0*}(896)$ substates $(m_z = \pm 1, 0)$ and of $K + \pi$ decay products in B field.

It is obvious that $K^{0*} \to K^- + \pi^+$ decay is kinematically forbidden in field $B > 1.5 \cdot 10^{15}$ T since the energy of (K^-, π^+) decay products exceeds the mass of K^{0*} meson. In such case the isospin conservation rule (leading to 2:1 ratio of decay probabilities) becomes violated, exactly as it happens with $D^{0*} \to D\pi$ strong decays in vacuum $(B = 0)$.

Let us have a look what happens in the case of charged K^* meson decays. Magnetic moment of $K^{\pm*}$ mesons $\mu_{K^{\pm*}} = \pm(\mu_u + \mu_{\bar{s}}) = \pm(1.85 + 0.61) = \pm 2.46\mu_N$ is 7x larger compared to $\mu_{K^{0*}}$ case and the additional contribution from $n = 0$ Landau level modifies the energy of $K^{\pm*}$ mesons. In the "small" field limit and static $(p_z = 0)$ case one can write

$$E[B, J_z] = m_{K^*} + |q_{K^*}|B/2m_{K^*} - J_z\mu_{K^*}B \tag{6}$$

which is sufficient for our purposes here. From Figure 4 one can conclude that both decay channels ($K^{+*} \to K^+ + \pi^0$ and $K^{+*} \to K^0 + \pi^+$) are affected and that the isospin violation effects are going to be much weaker if compared to the case of K^{0*} decays.

If K^{0*} and $K^{\pm*}$ resonance yields in ultra-relativistic heavy ion collisions are determined from their $K^\pm + \pi^\mp$ and $K^0_s + \pi^\pm$ decays, and strong magnetic field effects we discuss here are really significant, one can expect that K^{0*}, $K^{\pm*}$ yields (determined with the assumption of isospin conservation in strong decays) will be different. This happens because of different degree of the isospin violation we predict here for K^{0*} and $K^{\pm*}$ mesons in the external magnetic field.

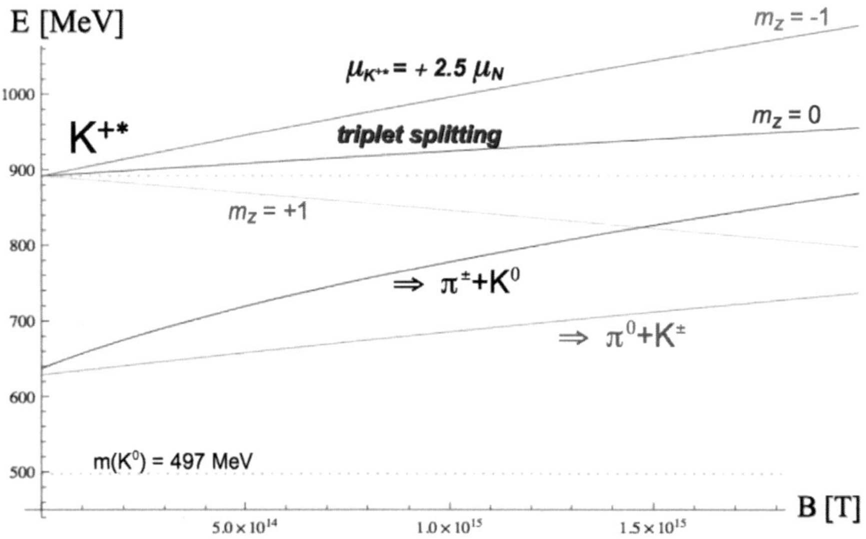

Figure 4. Energy of $K^{+*}(892)$ substates $(\pm1, 0)$ and of $K + \pi$ decay products in B field.

Additionally, from Figure 4 one can conclude that different J_z spin projections of $K^{\pm*}(892)$ mesons are affected differently by the restricted phase-space available for $K^0_s + \pi^\pm$ decays, which may consequently lead to the measurable tensor polarization of $K^{\pm*}$ mesons.

6. Decay of Λ^* and Ξ^{0*} in strong magnetic field

Energy of neutral baryons in the magnetic field is described by Eq.(3) which leads to quadruplet energy splitting for $J = 3/2$ case. Magnetic moment of $\Xi^{0*}(1530)$ resonance can be estimated as $\mu_{\Xi^{0*}} = 2\mu_s + \mu_u = 0.62\mu_N$ using the constituent quark model for $u^\uparrow s^\uparrow s^\uparrow$ quarks with parallel spins in $L = 0$ quantum state, while for $\Lambda^*(1520)$ we shall use value $\mu_{\Lambda^*} = -0.2\mu_N$ [10].

Figure 5 suggests that $\Xi^{0*} \to \Xi^- + \pi^+$ decay channel becomes energetically forbidden in field $B > 4 \cdot 10^{14}$ T for both orientations of Ξ^- spin $s_z = \pm 1/2$, while strong decays $\Xi^{0*} \to \Xi^0 + \pi^0$ remain kinematically allowed. At even higher field strengths the interaction of magnetic moment $\mu_{\Xi^0} = 1.25\mu_N$ with the external field (for $s_z = -1/2$) may bring the energy of $\Xi^0 + \pi^0$ pair above the mass of Ξ^{0*} substates, but this is not our concern here. Our main conclusion is, that due to the restricted phase space for $\Xi^{0*} \to \Xi^- + \pi^+$ decays the isospin conservation in decays $\Xi^{0*} \to \Xi + \pi$ can become violated in a sufficiently strong (static) magnetic fields.

Lifetime of Ξ^{0*} baryon state ($\approx$ 21 fm/c) is considerably longer compared to the expected magnetic field duration in relativistic heavy ion collisions, and therefore we are reluctant to relate the smaller yield of Ξ^{0*} baryons discussed at RHIC [11] with the magnetic field-induced phenomena we study here. However, lifetime $\tau_{\Lambda^*} \approx 13$ fm/c of resonance $\Lambda^*(1520)$ is shorter, and therefore the behavior of Λ^* in the magnetic field may be of some relevance.

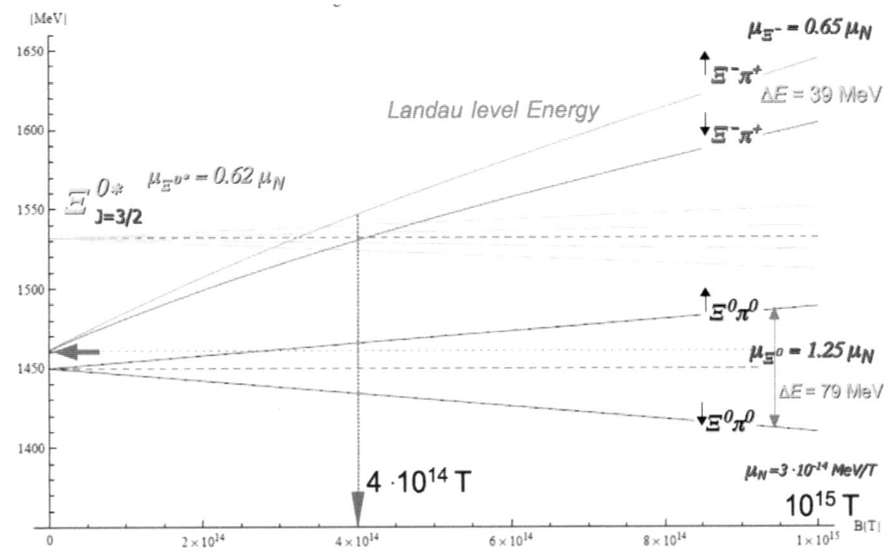

Figure 5. Energy of $\Xi^{0*}(1532)$ baryon and of its decay products $\Xi^-\pi^+$ and $\Xi^0\pi^0$ in static magnetic field (using $\mu_{\Xi^{0*}} = 2\mu_s + \mu_u = 0.62\mu_N$ and $s_z = \pm 1/2$ orientations for Ξ^- and Ξ^0).

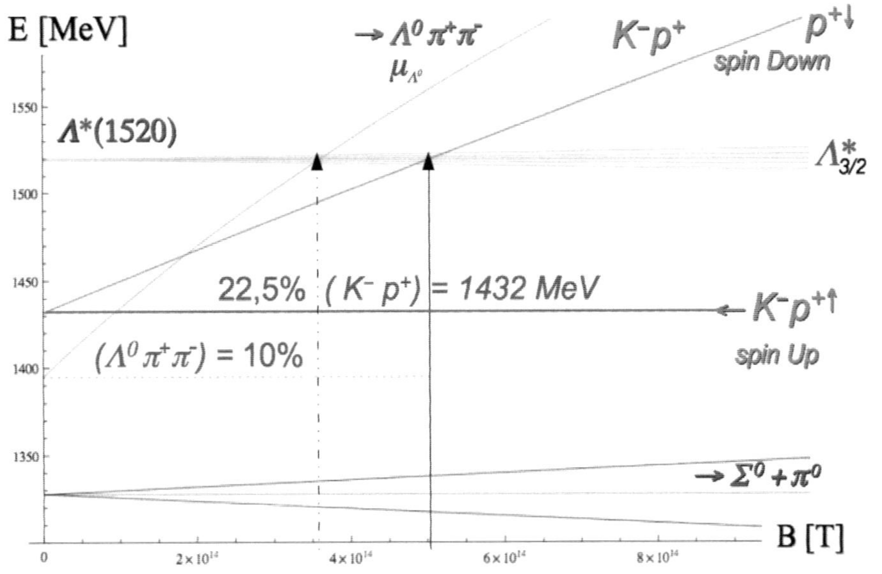

Figure 6. Energy of $\Lambda^*_{3/2}(1520)$ baryon (s_z=$\pm 1/2$,$\pm 3/2$) substates (using μ_{Λ^*}=$-0.2\mu_N$) and of its decay products K^-p^+, $\Lambda^0\pi^+\pi^-$ and $\Sigma^0\pi^0$ in the external static magnetic field.

In Figure 6 we show that decay channel $\Lambda^* \to \Lambda^0\pi^+\pi^-$ becomes kinematically closed in field $B > 3.5 \cdot 10^{14}$ T, while channel $\Lambda^* \to \Sigma^0\pi^0$ remains open in even much higher magnetic fields.

For decay $\Lambda^* \rightarrow \Lambda^0\pi^+\pi^-$ we do not show Λ^0 energy splitting due to $\mu_{\Lambda^0} = -0.613\mu_N$ interaction with B field, because the effect is small, and this channel is not relevant here. However, for $\Lambda^* \rightarrow p + K^-$ decay channel, which is used to reconstruct $\Lambda^*(1520)$ in the experiments, the interaction $\Delta E = -\vec{B} \cdot \vec{\mu}_p$ is important.

For $s_z = -1/2$ proton spin projection, the energy of $(p + K^-)$ pair rises above Λ^* mass at $B > 5 \cdot 10^{14}$ T. This happens because $n = 0$ Landau energy increase of K^- meson and Proton is supplemented by $\Delta E = +\mu_p B$ positive energy contribution. However, in $p^{\uparrow}(s_z = +1/2)$ case, the interaction of $\mu_p = 2.79\mu_N$ with the field just compensates the rising contribution from $n = 0$ Landau energy of charged K and p particles, what gives almost horizontal line at $E = 1432$ MeV if Figure 6. Thus 50% of $\Lambda^* \rightarrow p + K^-$ decays become suppressed in $B > 5 \cdot 10^{14}$ T field, while $\Lambda^* \rightarrow n + \bar{K}^0$ channel remains open and (almost) unaffected.

Precise measurement of $\Lambda^*(1520)$ production in heavy ion collisions might be interesting, because Λ^* signal observed in peripheral and central Au+Au collisions at RHIC [12] is missing in non-central Au+Au collisions [13]. It remains to be resolved whether rescattering mechanism or just statistics, or decay channel suppression phenomenona discussed here are responsible for the unobserved $\Lambda^*(1520)$ peak in non-central Au+Au collisions at RHIC.

Decays of $\Delta(1232)$ resonance with its short lifetime $\tau \approx 2$ fm/c may also be influenced by magnetic fields of strength $B \approx 5 \cdot 10^{14}$ T. Interested readers may find related information in [14].

7. Enhanced CP violation in η_c and $\eta(547)$ decays due to magnetic field

In the preceding sections we have studied the interaction of magnetic field with charge and magnetic moments of hadrons. We have found that isospin conservation rules in strong decays of resonances may become violated due to kinematical (energy conservation) and phase-space reasons. However, the interaction of constituent quark magnetic moments with external magnetic field may change also the internal structure of hadrons and their decay properties, as we shall describe here.

It has been mentioned already by Gell-Mann and Pais [16] that rigorous conservation of C parity should be expected only in the absence of external fields. Indeed, ortho-Positronium (o-Ps) $s_z = 0$ substate (which may decay only via o-Ps $\rightarrow 3\gamma, 5\gamma$ channels in vacuum) obtains quantum admixture of para-Ps ($J = 0$) state in the magnetic field. In such situation, (e^+e^-) bound state with $J = 1, s_z = 0$ quantum numbers can decay into $\rightarrow \gamma\gamma$ pairs, which leads to experimentally observed [15] "magnetic field quenching" of ortho-Ps $\rightarrow 3\gamma$ decays. The phenomenon occurs because new eigenstates Ψ_o^+ and Ψ_p^- of the Hamiltonian containing the interaction of μ_{e^+} and μ_{e^-} magnetic moments with external magnetic field are (for $B \neq 0$) superpositions of the original C parity eigenstates: $\Psi_p(J = 0) = (\uparrow\downarrow - \downarrow\uparrow)/\sqrt{2}$ and $\Psi_o(J = 1, s_z = 0) = (\uparrow\downarrow + \downarrow\uparrow)/\sqrt{2}$:

$$\Psi_o^+ = \cos(\alpha)\Psi_o + \sin(\alpha)\Psi_p \qquad \Psi_p^- = \cos(\alpha)\Psi_p - \sin(\alpha)\Psi_o \tag{7}$$

Mixing angle α depends on the external magnetic field B as $\sin(\alpha) = y/\sqrt{1 + y^2}$, where $y = x/(1 + \sqrt{1 + x^2})$ and $x = 4\mu_e B/\Delta E_{hf}$ (here $\Delta E_{hf} = 8.4 \cdot 10^{-4}$eV is hyperfine energy splitting of Positronium [8]). Since constituent quarks have their magnetic moments (one may expect $\mu_c \approx 0.4\mu_N$ and $\mu_b \approx -0.07\mu_N$), eigenstates $\Psi(nS)$ and $\Upsilon(nS)$ of $J = 1$ Quarkonium mesons should be affected [5] by external magnetic fields in a similar way as ortho-Positronium. Hyperfine energy splitting between $(J = 0)$ $\eta_{c,b}$ mesons and $(J = 1)$ Ψ, Υ states is 116MeV and 71MeV, and the required magnetic field for the mixing effect to occur in Quarkonium is [5]: $B \approx 10^{14} - 10^{15}$ T. This phenomenon (suggested at 2012 DSPIN conference [17]) allows mesons, in quantum superposition state given by Eq.(7), to decay via new channels (in magnetic field). For example, $s_z = 0$ substate of $\Upsilon(1s)$ meson (decaying to $\rightarrow ggg$ channel in 82% of cases) may disintegrate via much faster $\rightarrow 2g$ (gluons) channel in the magnetic field, which can shorten the lifetime of Υ mesons considerably [5] and also suppress the amount of $\Upsilon \rightarrow l^+l^-$ decays.

However, in this contribution we would like to discuss another consequence of the above mentioned quantum mixing phenomenon: In the magnetic field, para-Positronium $J = 0$ state acquires admixture of o-Ps($J = 1, s_z = 0$) substate. This is not very interesting in Positronium case, because lifetime ratio $\tau_{o-Ps}/\tau_{p-Ps} \approx 10^3$ results in very small influence [8] of the magnetic field on para-Ps decays (possibility to decay via $\rightarrow 3\gamma$ channel does not influence significantly quantum state which can decay via $\rightarrow 2\gamma$ "fast" channel by default).

However, in the case of $\eta(547)$ and $\omega(782)$ mesons, the ratio of lifetimes is opposite $\tau_\omega/\tau_\eta = 1.5 \cdot 10^{-4}$, which means that $r = 10^{-6}$ admixture of ω meson in $\Psi_p^- = \tilde{\eta}(547)$ superposition ($J = 0$) state introduces $\approx 0.6\%$ contamination from ω decays [19]. This allows $\tilde{\eta}(547)$ meson in the magnetic field to decay also into $\rightarrow \pi^+\pi^-$ final states, which is otherwise forbidden by CP conservation at the level BR($\eta \rightarrow \pi^+\pi^- \leq 10^{-27}$) within Standard Model. For $\eta_c - J/\Psi$ mixing, the enhancement of $\eta_c \rightarrow \pi^+\pi^-$ decays is much less significant ($\tau_\Psi/\tau_{\eta_c} \gg 1$).

It seems that magnetic field is able to enhance CP - violating decays of bound hadronic systems via quantum mixing - superposition phenomenon. This may occur only if "G" parity violation is significant (J/Ψ, $\omega(782) \rightarrow \pi^+\pi^-$ decays cannot occur when G parity is conserved).

8. Superposition of J/Ψ, Υ with $\chi_{c,b}$ states in the external field

When Quarkonium vector ($J=1$) mesons move in the magnetic field, they experience (in their rest frame) electric field $E_t = -\gamma v_t B_z$, which can influence their internal quantum state in a specific way. For Positronium, such phenomenon (motional Stark effect) has been studied carefully [4]. It has been established that $s_z = \pm 1$ states of ortho-Positronium acquire admixture from $1P$ (e^+e^-) states (these correspond to $\chi_c(J = 0, 1, 2)$ mesons in the case of Charmonium). For example, eigenstate $\tilde{\Psi}^{s_z}(c\bar{c})$ with $J=1, L=0, s_z=+1$ quantum numbers of the Hamiltonian containing the interaction term for external electric fields is

$$\tilde{\Psi}^{+1}(2s) = \Psi^{+1}(2s) - x_2(\frac{\chi_{c2}^0}{\sqrt{6}} - \chi_{c2}^{+2}) - x_1\frac{\chi_{c1}^0}{\sqrt{2}} + x_0\frac{\chi_{c0}^0}{\sqrt{3}} \qquad (8)$$

where parameters x_2, x_1, x_0 depend on "hyperfine" energy splittings ΔE_{hf} between $\Psi(2s)$ and $\chi_{c0}, \chi_{c1}, \chi_{c2}$ mesons. One has $x_2 = 3\sqrt{2}\, q_c\langle r_0\rangle E_t/\Delta E_{hf}(\Psi_{2s} - \chi_{c2}) = 2\sqrt{2}\, e\langle r_0\rangle E_t/130\text{MeV}$, where $\langle r_0\rangle = 0.45$fm is the "size" of $\Psi(2s)$ state. Expressions for x_1 and x_0 are the same as x_2, just ΔE_{hf} parameters are 175MeV and 271MeV for ($\Psi_{2s} - \chi_{c1}$) and ($\Psi_{2s} - \chi_{c0}$) mass differences.

We mention this here to point out, that $\Psi(nS) - \chi_c(nP)$ and $\Upsilon(nS) - \chi_b(\tilde{n}P)$ superpositions (in external electric field) are more significant in case of small ΔE_{hf} mass splittings between $\chi_{c,b}$ and triplet $J=1$ Quarkonium states. From the known masses of Charmonium and Bottomium mesons one can find that $\Psi(2s)$ and $\Upsilon(2s, 3s)$ states are more affected by quantum mixing effects (in the magnetic and electric field) than $J/\Psi(1s)$ and $\Upsilon(1s)$ mesons.

Admixtures of $\chi_{c,b}$ and $\eta_{c,b}$ states in $\Upsilon(ns)$ and $\Psi(ns)$ mesons allow them to decay via "fast" ($\rightarrow gg$) channels, lowering the yield of l^+l^- pairs from their decays. We suggest that suppression/disappearance of $\Upsilon(2s, 3s)$ yields observed in Pb+Pb collisions at LHC [18] might possibly be related also to the quantum mixing behavior we have discussed here.

9. Summary

We have studied the influence of external magnetic and motional electric field on decay properties of mesons and baryonic resonances. We suggest that isospin conservation rules determining branching ratios for various strong decay channels of K^*, Ξ^*, Λ^* hadronic resonances may become violated in the magnetic field ($B \approx 10^{14}$ T) due to the restricted phase-space and energy conservation. In the case of $\rho^0 \rightarrow \pi^+\pi^-$ decay suppression, we speculate that enhanced photon and dilepton production may occur. CP violation in $\eta \rightarrow \pi^+\pi^-$ decay channel has been predicted to be enlarged in the magnetic field, and the influence of motional electric field on $\Psi(ns)$ and $\Upsilon(ns)$ Quarkonium states has been considered.

Acknowledgments

This work was performed within project No. 2/0197/14 of the Slovak Grant Agency VEGA. Financial support from Slovak Research and Development Agency project No. APVV-0177-11 is acknowledged. The author thanks to Slovak CERN committee for the encouraging support and to K. Šafařík for the kind hospitality at CERN, where some of ideas presented here had been elaborated. The invitation to give a talk at WWND 2015 conference and excellent management of this scientific meeting (by R. Bellwied, F. Geurts, and A. Timmins) are highly appreciated.

References

[1] Chernodub M N 2011 *Physical Review* **D82** 085011
[2] Deng W-T and Huang X-G 2012 *Physical Review* **C85** 044907
[3] Hughes V W, Marder S and Wu C S 1955 *Physical Review* **98** 1840
[4] Dermer C D and Weisheit J C 1989 *Physical Review* **A40** 5526
[5] Filip P 2013 *Proceedings of Science, PoS (CPOD 2013)* **035**
[6] Landau L D and Lifschitz E M 1977 *Course of Theoretical Physics* Vol. **3** (London: Pergamon Press)
[7] Suh I-S and Mathews G J 2001 *The Astrophysical Journal* **546** 1126
[8] Berko S and Pendleton H N 1980 *Annual Review of Nuclear and Particle Science* **30** 543
[9] Adare A *et al.* (PHENIX Collaboration) 2012 *Physical Review Letters* **109** 122302
[10] Torres A M, Khemchandani K P, Sharma N and Dahiya H 2012 *European Physics Journal* **A48** 185
[11] Kerbikov B O and Malinina L V 2010 *Physical Review* **C81** 034901
[12] Abelev B I *et al.* (STAR Collaboration) 2006 *Physical Review Letters* **97** 132301
[13] Gaudichet L 2004 *Journal of Physics* **G**: *Nucl. Part. Phys.* **30** S549
[14] Filip P 2014 *Proceedings of Science, PoS (Baldin ISHEPP XXII)* **119**
[15] Hughes V W, Marder S and Wu C S 1955 *Physical Review* **98** 1840
[16] Gell-Mann M and Pais A 1955 *Physical Review* **97** 1387
[17] Filip P 2014 *Physics of Particles and Nuclei* **45** 7
[18] Chatrchyan S *et al.* (CMS Collaboration) 2011 *Physical Review Letters* **107** 052302
[19] Filip P 2014 *EPJ Web of Conferences* **81** 05013

Quark coalescence from RHIC to LHC

Vincenzo Minissale[1,2], Francesco Scardina[1,2] and Vincenzo Greco[1,2]

[1] Department of Physics and Astronomy, University of Catania, Via S. Sofia 62, I-95123, Catania, Italy.

[2] INFN-LNS,Laboratori Nazionali del Sud, Via S. Sofia 64,I-95123, Catania, Italy

E-mail: vincenzo.minissale@lns.infn.it, scardinaf@lns.infn.it, greco@lns.infn.it

Abstract. We discuss the application of a phase-space coalescence plus fragmentation model for the hadronization of the quark-gluon plasma (QGP) created in ultra-relativistic heavy-ion collisions. Recombination of minijet partons with the partons from the QGP is also included and plays a role at $p_T \sim 2-4$ GeV where the baryon to meson anomaly is observed experimentally. We show our prediction for light and strange hadrons transverse momentum spectra (π, p, k, Λ) and baryon to meson ratios (p/π, Λ/k) in a wide range of p_T, both for RHIC and LHC energies. The baryon to meson ratio at LHC presents similar features of that at RHIC, but with a shift in the peak of about 0.5 GeV, and this is well predicted by our model.

1. Introduction

Heavy Ion Collisions (HIC) at ultrarelativistic energy can be used to probe the properties of nuclear matter under such extreme conditions. The predictions of Lattice Quantum Chromodynamics indicate that the critical temperature in which the nuclear matter experiences a phase transition is $T_c \approx 160$ MeV, and new state of matter can exist in which the colour charges are deconfined in a Quark Gluon Plasma (QGP) [1]. Signatures of a Quark Gluon Plasma formation became manifest in the experiment with energies up to 200 AGeV performed at RHIC and further confirmations as well as new discoveries have been coming from the experiments at LHC. In the studies of the QGP created in HIC is necessary taking into account that partonic behaviour is not directly projected on the observables measured in heavy ion collision experiments. Thus the choice of the model for hadronization process is a crucial point in order to have a comparison with experimental data. Some unexpected observations in Au+Au collisions at RHIC have been the enhancement of baryon to meson ratio at intermediate transverse momenta and the scaling of elliptic flow according to the constituent quark number. The coalescence hadronization process can explain both this features that appears in the same intermediate p_T region [2, 3, 4]. The coalescence approach born from the idea that comoving partons in the quark-gluon plasma combine their transverse momentum to produce a final-state meson or baryon with higher p_T than the partons themselves if they are in a bulk QGP matter at a time scale in which the partons can rescatter and generate a thermal medium expanding with a collective radial flow [10, 11, 5, 6, 7, 8, 9]. The hadron spectra at high transverse momentum are dominantly produced from minijet partons originated from initial hard processes between colliding nucleons. In the coalescence model used here, these minijet partons are allowed to recombine with thermal quarks and antiquarks from the quark-gluon plasma created in the collisions to form hadrons. We report in this paper results obtained using coalescence model for central collision (0-10%), $Au + Au$ at RHIC with $\sqrt{s}$=200 GeV and $Pb + Pb$ at LHC with $\sqrt{s}$=2.76 TeV.

2. Hadronization via Coalescence and Fragmentation

2.1. Fragmentation

The fragmentation is the dominant way to hadronize in ultra-relativistic proton-proton collisions. Fragmentation functions are not reliably calculable from first principles in QCD. However, they are observables and can be measured experimentally. In this case are used the Albino-Kniehl-Kramer (AKK) fragmentation function [12, 13].
As said before the hadron spectra at high transverse momentum are dominantly produced from minijet partons. The amount of energy loss of minijet partons is consistent with the scenario that they have traversed through a dense matter that consists of coloured quarks and gluons. The recombination between minijet partons and thermal partons from the quark-gluon plasma is important for production of hadrons with intermediate transverse momentum, leading to comparable antiproton and pion yields in this momentum region as observed experimentally. It further predicts that the antiproton to pion ratio would decrease as their transverse momenta become large. In this high transverse momentum region, independent fragmentations of minijet partons dominate particle production and lead to a very small antiproton to pion ratio.
The hadron spectra are given by:

$$\frac{dN_{had}}{d^2p_T\,dy} = \sum_{jet}\int dz \frac{dN_{jet}}{d^2p_T\,dy}\frac{D_{had/jet}(z,Q^2)}{z^2} \tag{1}$$

where D are the partons fragmentation functions, z is the fraction of minijet momentum carried by the hadron and $Q^2 = (p_{had}/2z)^2$ is the momentum scale for the fragmentation process.

2.2. Coalescence

The coalescence approach is based on a Wigner function formalism. The hadron spectrum can be written as:

$$\frac{dN_H}{d^2P_T} = g_H \int \prod_{i=1}^{n} \frac{d^3p_i}{(2\pi)^3E_i} p_i \cdot d\sigma_i \; f_q(x_i,p_i) f_H(x_i..x_n,p_i..p_n)\delta^{(2)}\left(P_T - \sum_{i=1}^{n} p_{T,i}\right) \tag{2}$$

where $f_H(x_i..x_n,p_i..p_n)$ is the Wigner distribution function and is the probability for n quark to form an hadron, $d\sigma$ denotes an element of a space-like hypersurface, g_H is the probability of forming a color neutral object with the spin of the hadron considered from n coloured quarks.The function $f_q(x,p)$ is the covariant distribution functions of quarks (and antiquarks) in the phase space. The light hadrons Wigner function that is used is a sphere in both space and momentum, with radii Δ_r and Δ_p, respectively, which in the Wigner formalism are related by $\Delta_r \cdot \Delta_p = 1$. For example, the meson wigner function is:

$$f_M(x_1,x_2;p_1,p_2) = \frac{9\pi}{2(\Delta_x\Delta_p)^3}\,\Theta(\Delta_x^2 - (x_1-x_2)^2)\Theta(\Delta_p^2 - (p_1-p_2)^2 + (m_1-m_2)^2) \tag{3}$$

A good description of pion, kaon, proton, antiproton and Λ spectra can be obtained with a radius parameter Δ_p = 0.19 GeV for mesons, 0.33 GeV for proton, 0.38 GeV for Lambdas, which in terms of mean square radius corresponds to take a slightly smaller radius for Λ as one can expect from the harmonic oscillator wave function, which has a width that scales with the inverse square root of the system reduced mass.

These multidimensional integrals are evaluated in the full 6D phase space by the Monte-Carlo method via test particle. In order to well describe the particle spectra, have been included some resonances decays. Is also taken in account the Boltzmann probability to produce an excited state with the statistical factor $exp(-(E_{H^*} - E_H)/T)$, where $E_{H^*} = (p_T + m_{H^*})^{1/2}$. Are also taken in account the degeneracy factors that come from different values of isospin and total angular momentum.

- π (I=1, J=0)
  - k^* (I=1, J=1/2); $k^* \longrightarrow k\pi$
  - ρ (I=1, J=1); $\rho \longrightarrow \pi\pi$
  - Δ (I=3/2, J=3/2); $\Delta \longrightarrow N\pi$
- proton (I=1/2, J=1/2)
  - Δ (I=3/2, J=3/2); $\Delta \longrightarrow N\pi$
- $k^{\pm}$ (I=0, J=1/2)
  - k^* (I=1, J=1/2); $k^* \longrightarrow k\pi$
- $\Lambda(1116)$ (I=0, J=1/2);
  - $\Sigma_0(1193)$ (I=1, J=1/2); $\Sigma_0 \longrightarrow \Lambda\gamma$
  - $\Lambda(1405)$ (I=0, J=1/2); $\Lambda(1405) \longrightarrow \Sigma\pi$
  - $\Sigma(1385)$ (I=1, J=3/2);
    - $\Sigma(1385) \longrightarrow \Lambda\pi$ with $B.R. =$ 88%
    - $\Sigma(1385) \longrightarrow \Sigma\pi$ with $B.R. =$ 11.7%

3. Fireball parameters

In this study is considered a fireball of thermalized particles which includes light quarks, antiquarks and gluons at a temperature corresponding to $T_c = 165$ MeV, that is about the cross-over transition temperature [1, 14]. Partons position in transverse direction has a uniform distribution. Longitudinal positions are distributed according to $z = \tau \sinh y$, assuming Bjorken correlation $\eta = y$. Assuming a constant acceleration, we connect the radial expansion $R_{\perp} = R_0 + 0.5\beta_{max}\tau$ with the radial flow β_{max}. Both $R_{\perp}$ and the volume $V = \pi R_{\perp}^2 \tau$ could be in general considered as parameters that are evaluated taking in account the typical value of lifetime (τ) of the QGP. In such manner the radial flow and the volume has constraints due to the experimental value of total multiplicity dN/dy and total transverse energy dE_T/dy. For a radial uniform expansion, the values obtained for this quantities are: $\beta_{max} = 0.37$, $R_{\perp} = 8.7$ fm, $\tau = 4.5\,\mathrm{fm}/c$ at RHIC , and $\beta_{max} = 0.60$, $R_{\perp} = 10.2$ fm, $\tau = 7.8\,\mathrm{fm}/c$ at LHC. We notice that such a values correspond in one unity of rapidity to a volume of $V \sim 1100$ fm^3 at RHIC, while at LHC $V \sim 2500$ fm^3, which means an increase of a bit more than a factor of two in agreement with the estimate from pion HBT interferometry [15]. For their longitudinal momentum distribution we assume boost-invariance, i.e. a uniform rapidity distribution in the rapidity range $y \in (-0.5, +0.5)$.

For partons in the quark-gluon plasma we take a thermal distribution for transverse momenta up to $p_0 = 2$ GeV, in particular the momentum spectra for quark and antiquarks is given by:

$$\frac{dNq,\bar{q}}{d^2r_T\, d^2p_T} = \frac{g_{q,\bar{q}}\tau m_T}{(2\pi)^3} \exp\left(-\frac{\gamma_T(m_T - p_T \cdot v_T \mp \mu_q)}{T}\right) \tag{4}$$

where $g_q = g_{\bar{q}} = 6$ are the spin-color degeneracy of light quarks and antiquarks, and the minus and plus signs are for quarks and antiquarks, respectively. The slope parameter T is taken to be $T = 165\ MeV$. Masses of light quarks and antiquarks are taken to be $m_{u,d,\bar{u},\bar{d}} = 300\ MeV$ and $m_{s,\bar{s}} = 475\ GeV$, similar to the masses of constituent quarks. For the quark chemical potential μ_q, we use a value of light antiquark to quark ratio which lead to the antiproton to proton ratio observed at midrapidity in heavy ion collisions experiments.

Partons at high transverse momenta (greater than 2 GeV), as said, are mainly from the minijets produced in initial hard collisions among nucleons. These spectra can be parametrized (Table 1

Table 1. Parameters for minijet parton distributions at midrapidity from Au+Au at $\sqrt{s} = 200$ GeV

	A[$1/GeV^2$]	B[GeV]	n
g	$3.18 \cdot 10^4$	0.5	7.11
u, d	$9.79 \cdot 10^3$	0.5	6.84
$\bar{u}, \bar{d}$	$1.89 \cdot 10^4$	0.5	7.59
s	$6.51 \cdot 10^3$	0.5	7.36
$\bar{s}$	$8.02 \cdot 10^3$	0.5	7.57

Table 2. Parameters for minijet parton distributions at midrapidity from Pb-Pb at $\sqrt{s} = 2.76$ TeV

	A_1	A_2	A_3	A_4	A_5	A_6
g	23.46	4.84	8.08	2.78	2.79	2.31
quark	24.68	5.11	8.01	0.55	5.65	2.56

and Table 2) at RHIC (left equation) and at LHC (right equation) as

$$\frac{dN_{jet}}{d^2p_T} = A\left(\frac{B}{B+p_T}\right)^n \qquad \frac{dN_{jet}}{d^2p_T} = \frac{A_1}{\left[1+\left(\frac{p_T}{A_2}\right)^2\right]^{A_3}} + \frac{A_4}{\left[1+\left(\frac{p_T}{A_5}\right)^2\right]^{A_6}} \tag{5}$$

4. Spectra at RHIC and LHC

4.1. RHIC

We start with comparing the results obtained using the coalescence model with the spectra at RHIC in Au+Au collisions at $\sqrt{s} = 200\ GeV$.

In Fig.1 we show the transverse momentum spectrum of pions, we can see that an important

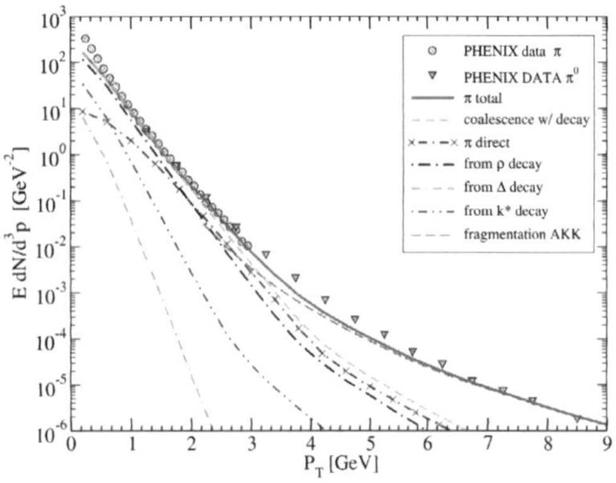

Figure 1. Pion transverse momentum spectrum at RHIC in Au+Au collisions at $\sqrt{s} = 200$ GeV. Experimental data from PHENIX [16] [17]

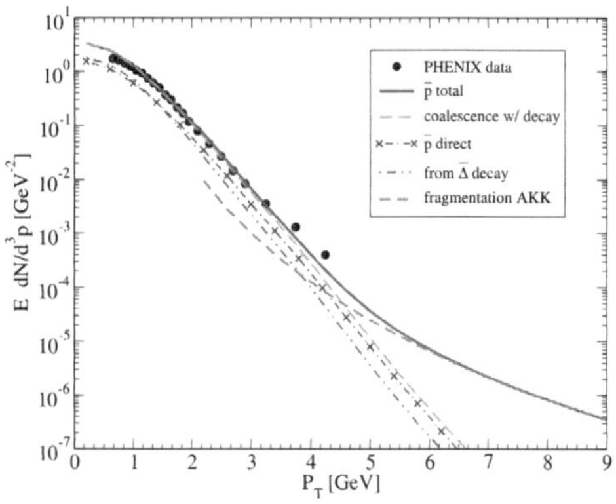

Figure 2. Antiproton transverse momentum spectrum at RHIC in Au+Au collisions at $\sqrt{s} = 200$ GeV. Experimental data from PHENIX [16].

contribution comes from the decays, in particular especially at low momenta the dominant one comes from ρ, in a region up to about $p_T \sim 3$ GeV, which is, also, the region anyway the fragmentation is starting to take over. The contribution from K^* and Δ are less relevant and only contribute to some little better description at very low momenta. It is known that for pions almost all the hadrons contribute to the feed-down, but in the region we are interested in, the resonances included are sufficient to have a good description for the spectra at $p_T < 1$ GeV.

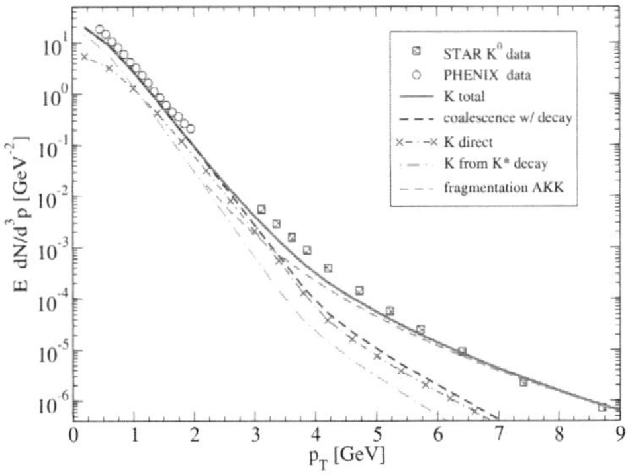

Figure 3. Kaon production at RHIC from Au+Au collisions at $\sqrt{s} = 200$ GeV. Experimental data from PHENIX [16], STAR [18].

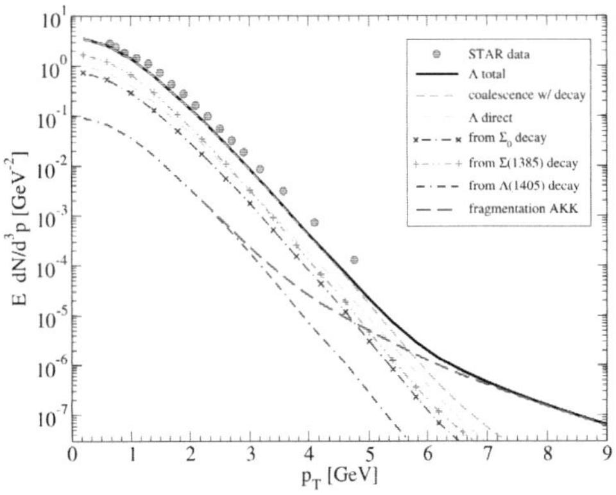

Figure 4. Λ production at RHIC from Au+Au collisions at $\sqrt{s} = 200$ GeV. Experimental data from STAR [19].

Can be seen that the production from coalescence and fragmentation becomes about similar for $p_T \sim 3.5$ GeV. In Fig.2 is shown the antiproton spectrum at RHIC. We show the relative contribution from coalescence and fragmentation by orange dashed line and brown dashed line respectively. We notice that for anti-protons the two mechanism become comparable at $p_T \sim 5$ GeV, that is a significant shift for protons with respect to pions. The description appears to be quite good. In Fig.3 we see that, also here, at low p_T the contribution from decay becomes important. K^* contribution is necessary in order to have a correct slope of the spectrum.

For the Λ, in Fig.4, there are indeed several hadronic state that have a significant contribution in the yield that comes from decays. We have included the closest resonances with their decay channel and the pertinent branching ratios (B.R.), as said above. Also for Lambdas the spectrum is correctly described in a wide range of p_T. We also find, similarly to the anti-proton, that the contribution from independent fragmentation becomes dominate at $p_T > 6$ GeV.

We can see in Fig.5 and Fig.6 that the ratios $\bar{p}/\pi^-$, Λ/K are well predicted, from the rise at

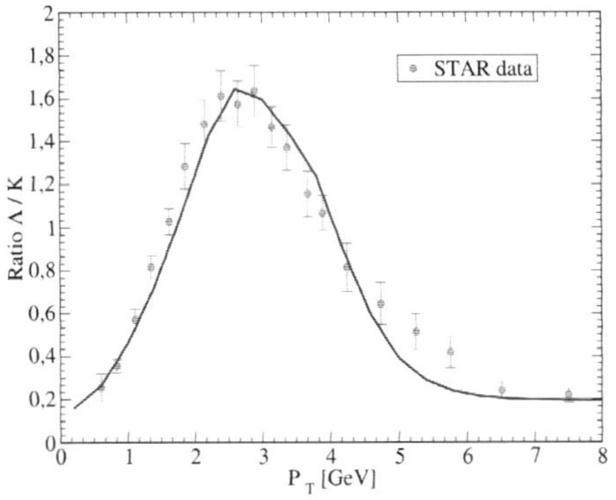

Figure 5. Lambda to kaon ratio at RHIC from Au+Au collisions at $\sqrt{s} = 200$ GeV. The model prediction is the solid line. STAR data by circles [20].

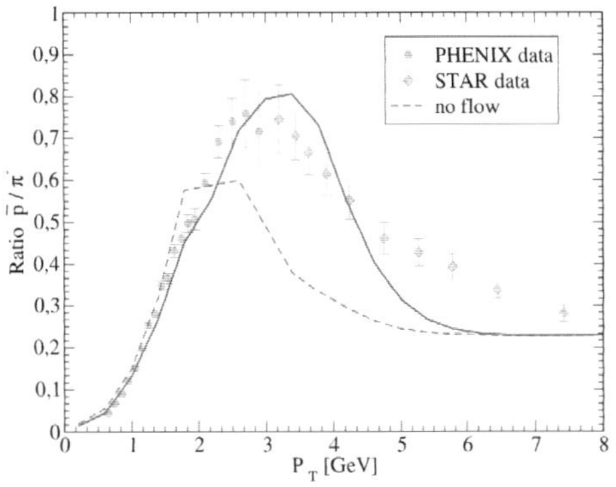

Figure 6. Particles ratio at RHIC from Au+Au collisions at $\sqrt{s} = 200$ GeV. Data from PHENIX [16] and STAR [18]

low p_T up to the peak region and then the falling-down behaviour. However in both cases it is clear that in the region of $p_T \sim 5-7$ GeV there is a lack of baryon yield. In Fig.6 is shown with the dashed line the result switching off the radial flow of the QGP matter. We can see how a coalescence model without radial flow could not reproduce the baryon over meson enhancement.

4.2. LHC

In Fig.7 is shown the transverse momentum spectrum for pions at LHC. The result obtained is quite in good agreement with the experimental data in all the transverse momentum region, except some lack of yield at very low p_T due to absence of all the resonance that are not taken in account. The yields from coalescence and fragmentation cross each other at $p_T \sim 4$ GeV, which is about a shift of 1 GeV with respect to RHIC. This is due to the larger collective flow present at LHC. Because the effect of radial flow is to shift the hadrons that are formed with coalescence mechanism, to larger p_T. Already here, we can see a very good agreement of the p_T distribution. The model is able to correctly predict the evolution, from RHIC to LHC, of the absolute yield and especially its p_T shape. The proton momentum spectrum is shown in Fig.8, also here there is a very good agreement is for $p_T > 1$ GeV up to 5 GeV. At very low p_T we have an overprediction for the yield. This discrepancy at LHC is more visible respect to RHIC case, and can be imputed to a larger expected annihilation of p and $\bar{p}$, this process could explain the disagreement between the statistical model prediction and the experimental data. The crossing between fragmentation and coalescence yields occurs at $p_T \sim 6$ GeV, which is about a momentum 50% larger with respect to the pions and is a shift of about 1.5 GeV with respect to RHIC. In Fig.9 the p_T distribution for K is shown by the red solid line. Also here there is good agreement with the experimental data in the entire range of p_T. We can notice that in the region where coalescence and fragmentation are comparable, at RHIC for both pions and kaons there was some lack of yield, instead at LHC energy for both cases the agreement appears quite better. The Λ spectrum at LHC, shown in Fig.10, is in good agreement for $p_T > 1$ GeV, in this case we have available data up to 9 GeV and this allows us to see that in the p_T region where the fragmentation dominates, i.e. $p_T \sim 6-7$ GeV, there is some lack of yield.

The baryon over meson ratios at LHC are shown in Fig.11 and Fig.12, there is a good agreement with experimental data, especially in the peak region. The model overestimate the ratio in the low p_T region, due to the production excess of p and Λ. These ratios show also that in the region where fragmentation becomes dominant, as anticipated before, the description is

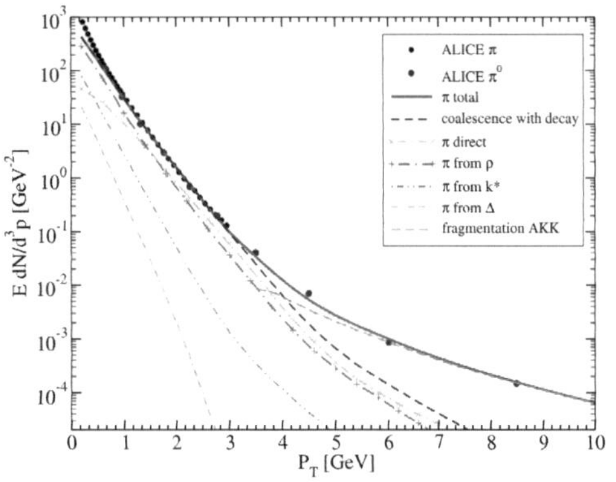

Figure 7. Pion transverse momentum spectrum from Pb+Pb collisions at $\sqrt{s} = 2.7\,TeV$. [21] [22].

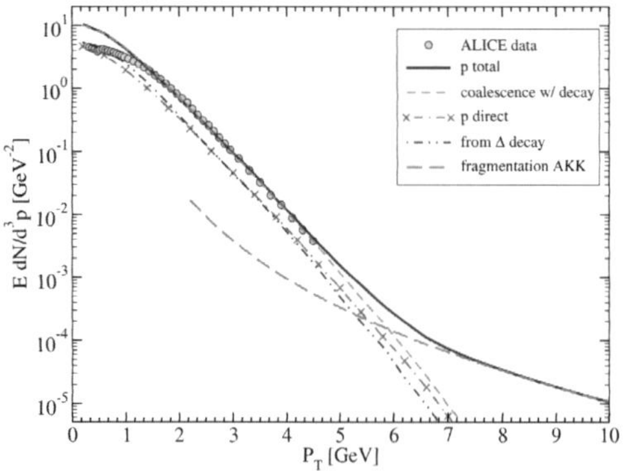

Figure 8. Proton transverse momentum spectrum from Pb+Pb collisions at $\sqrt{s} = 2.7\,TeV$. [21].

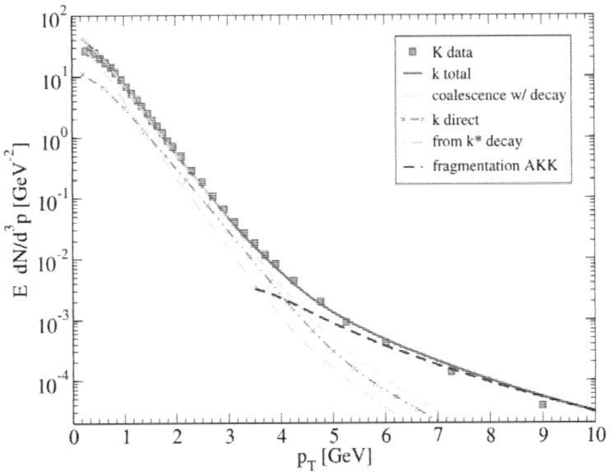

Figure 9. Kaon transverse momentum spectrum at LHC in Pb+Pb collisions at $\sqrt{s} = 2.7$ TeV.[21, 23, 24].

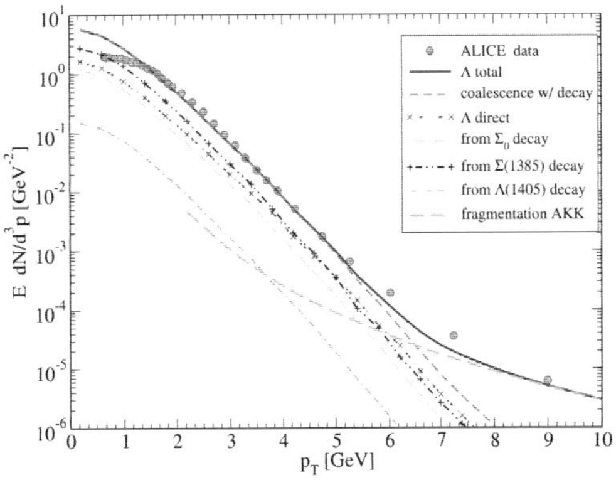

Figure 10. Λ transverse momentum spectrum at LHC in Pb+Pb collisions at $\sqrt{s} = 2.7$ TeV. [23] [24].

not so good. It is interesting to notice that recombination of thermal partons with a high-p_T minijet parton give an important contribution to reproduce well the ratio in the peak region, as shown by the black dashed curve in Fig.11 . Fragmentation function for baryons in general, and in particular for Λ are known not be very well constrained. In particular both at RHIC and at LHC seems that where in our approach coalescence becomes less important the spectrum from AKK fragmentation function appears too flat. Studies of in-medium fragmentation function can solve it or it could be that coalescence contribution should extend to large p_T with respect to the present modeling. There are studies that, recently, investigate a process that within the coalescence plus fragmentation approach could be quite important in solving this issue [26, 27]. They essentially describe the in-medium fragmentation as a quark recombination of shower partons taking into account also the gluon splitting in quark pairs that recombine. This mechanism can produce a large contribution for baryons in the region of $p_T \sim 6$ GeV.

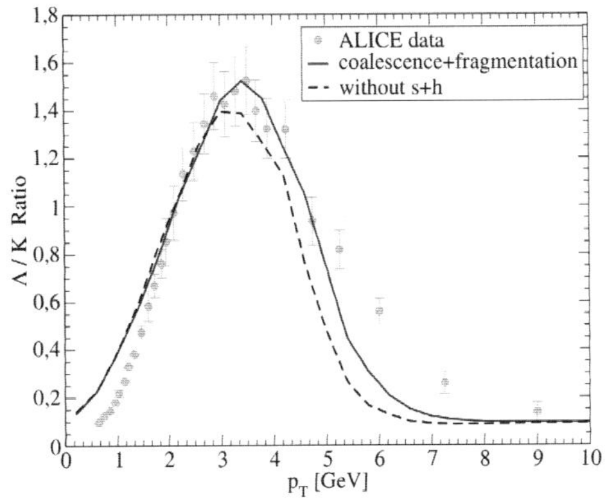

Figure 11. Lambda to kaon ratio in Pb+Pb collisions at $\sqrt{s} = 2.7$ TeV. [23]

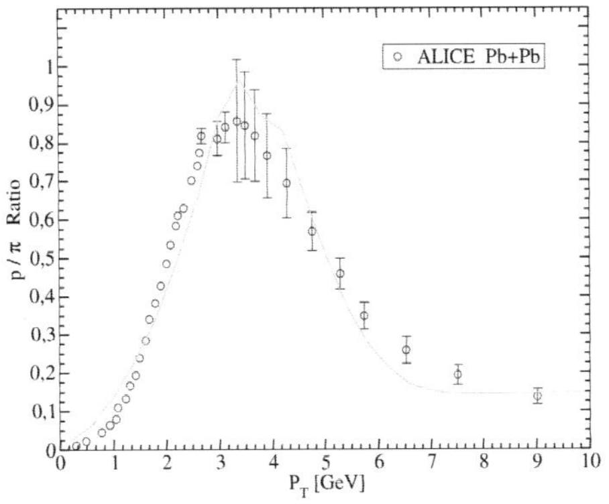

Figure 12. Proton to pion ratio in Pb+Pb collisions at $\sqrt{s} = 2.7$ TeV [25].

Journal of Physics: Conference Series **636** (2015) 012014 doi:10.1088/1742-6596/636/1/012014

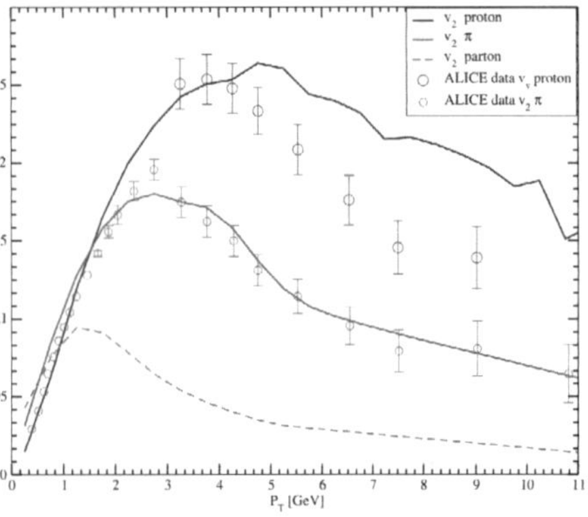

Figure 13. Elliptic flow of pions (green solid line) and protons (blue solid line) in function of transverse momentum in Pb+Pb collisions at $\sqrt{s} = 2.7\ TeV$. [28]

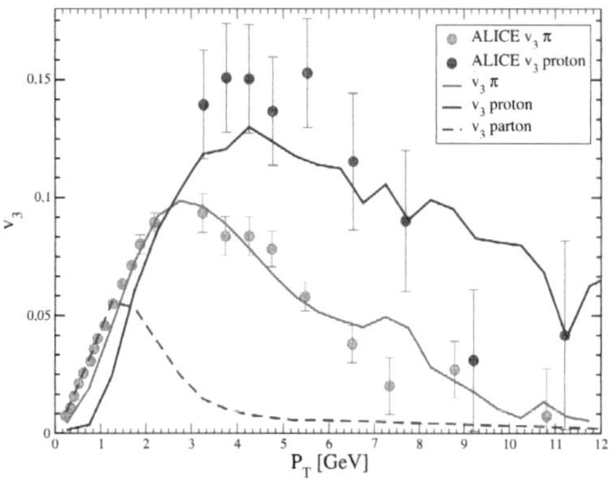

Figure 14. v_3 in function of transverse momentum in Pb+Pb collisions at $\sqrt{s} = 2.7\ TeV$. [28]

4.3. Elliptic flow

The azimuthal distribution of the particle transverse momentum spectra, i.e the probability of having a particle with a given transverse momentum p_T and an azimuthal angle ϕ, can be expressed in general in terms of the Fourier series $f(\phi, p_T) \propto 1 + 2\sum_{n=1}^{\infty} v_n(p_T) \cos n\phi$ where v_n denotes the n-th order momentum anisotropy, and v_2 coefficient is called elliptic flow. In coalescence model the meson (baryon) azimuthal distribution is given by the square (cubed) of azimuthal distribution of parton $f(\phi)$, and can be expressed in Fourier series.

In Fig.13, we show the parton elliptic flow by dashed line together with the pion elliptic flow shown by green solid line and the proton elliptic flow given by blue solid curve. These are compared to data measured in ALICE experiment, shown by circles. For proton elliptic flow the contribution from fragmentation has not been taken into account. We notice that the maximum pion v_2 is about a factor of two times larger than the quark one, while the proton v_2 is almost a factor of 3 larger, in fact the coalescence process propagates the anisotropy of constituent quark at partonic level to hadronic level according with $v_{2,M}(p_T) \approx 2\ v_{2,q}(p_T/2)$, $v_{2,B}(p_T) \approx 3\ v_{2,q}(p_T/3)$ so the proton elliptic flow is grater than pion elliptic flow by a factor $\sim 3/2$, and a similar shift of the p_T at which the v_2 reaches the maximum value. We can see, also, that the splitting between the pion and proton v_2 is well reproduced by the coalescence model.

It has been observed that due to local space fluctuation the angular distribution function of the hadron presents also a third-odd harmonics, $v_3 = \langle \cos 3\phi \rangle$. In Fig.14 parton v_3 relative to protons is shown by the dashed line while the v_3 relative to pion calculated in our model is given by the red solid line and v_3 for proton is indicated by the blue solid line. Experimental data by ALICE experiment are shown by circle, red for pions, blue for protons. We can see that both the large splitting in the v_3 of protons and pions as the location of the maximum are fairly well reproduced by the coalescence model.

5. Conclusions

The aim of this study has been to predict the evolution of the transverse momentum spectra from RHIC to LHC energy using the approach developed a decade ago used to investigate the RHIC data. In order to predict the feature at LHC no one adjustment of the coalescence parameter, like the width of Wigner function, has been made. The results obtained are in good agreement with the experimental data from LHC in a wide range of p_T. We also note that underlying such agreement a key role in the model is to ascribe to the radial flow of the QGP matter, a correct prediction of spectra and ratio is impossible without the effect due to radial flow. It appear that the independent fragmentation approach (at least using the AKK) gives too hard spectra at least up to $p_T \simeq 8$ GeV. The result seem to point to the need of an in-medium fragmentation process a results that is in agreement with the first results of a shower recombination of quarks from gluon decays. We remind that the coalescence has played an important role in the discussion about the QGP properties, in particular in the study of the anisotropy in the angular emission primarily measured by the so-called elliptic flow. Considering the elliptic flows as function of transverse momentum the coalescence model reproduces the observed splitting of pion and proton elliptic flow. We have extended the coalescence model to include also the next order harmonics of anisotropic flow, the v_3 coefficient. The predictions of our model for the v_3 harmonic present a splitting of pion and proton v_3 quite close to the one measured experimentally. A more realistic approach to coalescence in three dimension with radial flow correlations, finite hadronic wave function widths and resonance decays shows a relevant quark scaling breaking. At LHC appeared clearly that such a symmetry is broken by local space fluctuations but a quantitative approach requires an extension of the present approach to an event-by-event Monte Carlo approach.

[1] Y. Aoki, G. Endrodi, Z. Fodor, S. Katz, and K. Szabo, Nature **443**, 675 (2006), hep-lat/0611014.
[2] STAR Collaboration, J. Adams *et al.*, Phys.Rev. **C72**, 014904 (2005), nucl-ex/0409033.
[3] STAR Collaboration, J. Adams *et al.*, Phys.Rev.Lett. **95**, 122301 (2005), nucl-ex/0504022.
[4] PHENIX Collaboration, S. Adler *et al.*, Phys.Rev.Lett. **91**, 182301 (2003), nucl-ex/0305013.
[5] R. Fries, B. Muller, C. Nonaka, and S. Bass, Phys.Rev.Lett. **90**, 202303 (2003), nucl-th/0301087.
[6] V. Greco, C. Ko, and P. Levai, Phys.Rev.Lett. **90**, 202302 (2003), nucl-th/0301093.
[7] V. Greco, C. Ko, and P. Levai, Phys.Rev. **C68**, 034904 (2003), nucl-th/0305024.
[8] R. Fries, B. Muller, C. Nonaka, and S. Bass, Phys.Rev. **C68**, 044902 (2003), nucl-th/0306027.
[9] D. Molnar and S. A. Voloshin, Phys. Rev. Lett. **91**, 092301 (2003), nucl-th/0302014.
[10] T. Biro, P. Levai, and J. Zimanyi, Phys.Lett. **B347**, 6 (1995).
[11] T. Biro, P. Levai, and J. Zimanyi, Phys.Rev. **C59**, 1574 (1999), hep-ph/9807303.
[12] S. Albino, B. Kniehl, and G. Kramer, Nucl.Phys. **B725**, 181 (2005), hep-ph/0502188.
[13] S. Albino, B. Kniehl, and G. Kramer, Nucl.Phys. **B803**, 42 (2008), 0803.2768.
[14] S. Borsanyi *et al.*, JHEP **1011**, 077 (2010), 1007.2580.
[15] ALICE Collaboration, K. Aamodt *et al.*, Phys.Lett. **B696**, 328 (2011), 1012.4035.
[16] PHENIX Collaboration, S. Adler *et al.*, Phys.Rev. **C69**, 034909 (2004), nucl-ex/0307022.
[17] PHENIX Collaboration, S. Adler *et al.*, Phys.Rev.Lett. **91**, 072301 (2003), nucl-ex/0304022.
[18] STAR Collaboration, G. Agakishiev *et al.*, Phys.Rev.Lett. **108**, 072302 (2012), 1110.0579.
[19] STAR Collaboration, J. Adams *et al.*, Phys.Rev.Lett. **98**, 062301 (2007), nucl-ex/0606014.
[20] STAR Collaboration, G. Agakishiev *et al.*, Phys.Rev.Lett. **108**, 072301 (2012), 1107.2955.
[21] ALICE Collaboration, B. Abelev *et al.*, Phys.Rev.Lett. **109**, 252301 (2012), 1208.1974.
[22] ALICE Collaboration, Abelev, Betty Bezverkhny and others, Eur.Phys.J. C74 (2014) 3108,1405.3794.
[23] ALICE Collaboration,Abelev, Betty Bezverkhny and others, Phys.Rev.Lett. **111**, 222301 (2013),1307.5530.
[24] ALICE, D. Chinellato, J.Phys.Conf.Ser. **446**, 012055 (2013), 1211.7298.
[25] ALICE Collaboration, Abelev, Betty Bezverkhny and others, Phys.Lett **B736**, 196-207 (2014), 1401.1250
[26] K. C. Han, R. J. Fries, and C. M. Ko, J.Phys.Conf.Ser. **420**, 012044 (2013), 1209.1141.
[27] C. M. Ko, Talk at *3rd Non Equilibrium Dynamics and TURIC Workshop* (http://fias.uni-frankfurt.de/crete2014/talks/13/Ko.pdf), 13 June 2014, Crete.
[28] ALICE Collaboration, Abelev, Betty and others, Phys.Lett **B719**, 18-28 (2013), 1205.5761

Baryon stopping and Feynman scaling in the color glass condensate approach

F.O. Durães[1], A.V. Giannini[2], V.P. Gonçalves[3] and F.S. Navarra[2]

[1] Curso de Física, Escola de Engenharia, Universidade Presbiteriana Mackenzie, CEP 01302-907, São Paulo, Brazil
[2] Instituto de Física, Universidade de São Paulo, CEP 05315-970 São Paulo, SP, Brazil
[3] Instituto de Física e Matemática, Universidade Federal de Pelotas, CEP 96010-900, Pelotas, RS, Brazil

E-mail: duraes@mackenzie.br

E-mail: avgiannini@usp.br

E-mail: victorpbg@gmail.com

E-mail: navarra@if.usp.br

Abstract. In this work baryon stopping and leading baryon production are investigated in the context of the color glass condensate (CGC) formalism. We assume that at large energies the coherence of the projectile quarks is lost and that the leading baryon production mechanism changes from recombination to independent fragmentation. The phenomenological implications for net-baryon production in $pp/pA/AA$ collisions are analyzed and predictions for LHC energies are presented. We find that at very high energies the leading baryon x_F spectra become nearly energy independent and we reach the approximate Feynman scaling regime.

1. Introduction

In high energy hadronic collisions, baryons are produced both in the central and in the forward rapidity region. In the first case baryons are produced together with antibaryons and the net baryon number (baryons minus antibaryons) is small. In contrast, in the large rapidity region there are almost only baryons and no antibaryons. These experimental facts suggest that the forward baryons are produced from the valence quarks of the projectile, whereas low rapidity baryons are produced mainly from gluons and sea quarks.

1.1. Leading baryons

The outgoing baryons which have large longitudinal momentum ($x_F \geq 0.2$) and the same valence quarks (or at least one of them) as the incoming particles are called leading particles (LP).

The momentum spectra of leading particles have been measured already some time ago [1]. Recently, data on leading protons and neutrons produced in eletron-proton reactions at HERA with a c.m.s. energy one order of magnitude higher became available [2]. In the case of photoproduction data can be interpreted in terms of the Vector Dominance Model and can therefore be considered as data on LP production in vector meson-proton collisions. These measurements of LP spectra both in hadron-hadron and in eletron-proton collisions have renewed the interest on the subject, specially because the latter are measured at higher energies and therefore the energy dependence of the LP spectra can now be determined. It is important to have a very good understanding of these spectra for a number of reasons. They are the input for calculations of the LP spectra in hadron-nucleus collisions, which are a fundamental tool in the description of atmospheric cascades initiated by cosmic radiation. The precise knowledge of energy flow (LP spectra and inelasticity distributions) is necessary to interpret the results obtained in cosmic ray experiments, such as the Pierre Auger Collaboration.

In high energy heavy ion collisions at RHIC and LHC, the expression leading baryon spectrum is less used. Instead, we speak of baryon stopping to talk about essentially the same concept, namely the momentum (or rapidity) distribution of baryons emerging from valence quarks after the collision. As expected, the valence quarks are found at rapidities lower than the beam rapidities, and we say that they have been partially stopped. It is very important to know where the outgoing (leading) baryons are located in momentum space. If the stopping is large they will stay in the central rapidity region and affect the dynamics there, generating, for example, a baryon rich equation of state. The composition of dense matter is therefore relevant for the study of quark gluon plasma formation. In any case, before modelling p-A or A-A collisions one has to understand properly hadron-hadron processes. The LP spectra are also interesting for the study of diffractive reactions, which dominate the large x_F region. Since LP spectra are measured in reactions with low momentum transfer and go up to large x_F values, it is clear that the processes in question happen, at least partially, in the non-perturbative domain of QCD.

1.2. Perturbative QCD and gluon saturation

In the framework of the QCD parton model of high energy collisions, leading particles originate from the emerging fast partons of the collision debris. There is a large rapidity separation between fast partons and sea partons. Fast partons interact rarely with the surrounding wee partons. The interaction between the hadron projectile and the target is primarily through wee parton clouds. Fast partons might therefore filter through essentially unaltered. Based on these observations and aiming to study p-A collisons, the authors of Ref. [3] proposed a mechanism for LP production in which the LP spectrum is given by the convolution of the parton momentum distribution in the projectile hadron with its corresponding fragmentation function into a final leading hadron. This independent fragmentation scheme was, however, not supported by the then existing leading baryon spectra measurements [1]. It was shown that the essential ingredient to correctly reproduce the data was valence quark recombination [4]. Indeed, at lower energies ($\sqrt{s} \simeq 10 - 60$ GeV) what happens is rather a coalescence of valence quarks to form the LP and not an independent fragmentation of a quark or diquark to a nucleon. At higher energies this situation may change. Due to quantum evolution, the colliding systems become extremely dense and the average momentum transfer also increases. The incident proton constituents scatter, experience a large momentum transfer, the coherence of the projectile is destroyed completely and the scattered quarks and gluons fragment independently. As a consequence, the proton decays predominantly into a beam of "fast" mesons, with the baryon number shifted to small momentum fraction ($x_F \leq 0.1$). This mechanism of LP production implies a strong energy loss by the valence quarks and the consequent strong enhancement of the energy used to produce secondary particles, the so called inelasticity. In [5] this possibility was considered with the inclusion of gluon saturation effects. The authors pointed out that, while the change from valence quark recombination to independent fragmentation is a natural consequence of the hardening of the collisions, the precise energy at which this transition happens is determined by the saturation scale. In [6] it was shown that, assuming that valence quark recombination would still dominate LP production at very high energies, the role played by gluon saturation would be just to slow down the softening trend of LP spectra with increasing energies. It is however a very modest effect. The final conclusion was that, as long as quarks recombination (a non perturbative phenomenon) remains the dominant mechanism of LP production, the LP spectrum will not be a good place to search for gluon saturation effects.

A few years later, after the publication of the net-baryon rapidity distribution data by the BRAHMS Collaboration, the authors of [7] carried out an analysis of these data using a color glass condensate model in which valence quarks traverse the nuclear target, interacting and fragmenting independently into baryons. In a nuclear target the saturation scale is considerably larger than in a nucleon, even when the latter has much higher energies. They computed the rapidity distributions and transverse momentum spectra, obtaining an overall good agreement with data. At lower energies and larger rapidities they were not able to reproduce the p_T spectra. This could be an indication of the breakdown of independent fragmentation at lower energies. In [8] we repeated their calculation introducing a more reliable dipole scattering cross section. As a result we obtained a better fit of the spectra, but the discrepancy between the the model predictions and the low energy data persisted. In this contribution we discuss these calculations giving special attention to the LP spectra obtained with the same model and their behavior at very high energies.

2. Net-baryon production and the color glass condensate

As mentioned above, at higher energies new phenomena are expected to affect forward baryon production. This process requires the interaction of valence quarks with a relatively large momentum fraction (x_1) of the projectile with low momentum (x_2) partons in the target. In the low x regime the target is a dense system of partons (predominantly gluons) which may form the Color Glass Condensate (CGC), a state of very high partonic densities in which the nonlinear effects of QCD change the parton distributions and hence the cross sections. The CGC is characterized by a momentum scale (Q_s) which marks the onset of nonlinear (or saturation) effects and grows with the reaction energy. At increasing projectile energies the valence quarks receive a transverse momentum kick of the order of Q_s and hence above a certain energy the coherence of the projectile quarks is lost and the leading baryon production mechanism changes from recombination to independent fragmentation (see [8] and references therein).

In the CGC formalism the differential cross section of forward production of a hadron with transverse momentum p_T and mass m, at rapidity y, reads [8]:

$$\frac{dN}{d^2p_T dy} = \frac{1}{(2\pi)^2} \int_{x_F}^{1} \frac{dz}{z^2}\, D(z) \frac{1}{q_T^2}\, x_1\, q_v(x_1)\, \varphi\,(x_2, q_T)\ , \tag{1}$$

where $D(z) \equiv D_{B/q}(z) - D_{\bar{B}/q}(z)$ is the net-baryon fragmentation function, $z = E_B/E_q$ is the fraction of the energy of the fragmenting quark (E_q) taken by the emerging baryon B, the variable q_T is the quark transverse momentum and x_F represents the Feynman-x momentum of the produced baryon. Furthermore, $x_1\, q_v(x_1)$ is the valence quark distribution of the projectile hadron and the function $\varphi(x_2, q_T)$ is the unintegrated gluon distribution of the hadron target, which is given by:

$$\varphi(x_2, q_T) = 2\pi q_T^2 \int dr_T\, r_T\, \mathcal{N}(x_2, r_T) J_0(r_T q_T)\, , \tag{2}$$

where J_0 is a Bessel function and $\mathcal{N}(x_2, r_T)$ is the forward scattering amplitude of a color dipole of radius r_T off a hadron target.

The evolution of $\mathcal{N}(x_2, r_T)$ is described in the mean field approximation of the CGC formalism [9] by the BK equation [10]. This quantity encodes the information about the hadronic scattering and hence about the non-linear and quantum effects in the hadron wave function.

In general, it is assumed that it can be modelled through a simple Glauber-like formula, which reads

$$\mathcal{N}(x, r_T) = 1 - \exp\left[-\frac{1}{4}(r_T^2 Q_s^2)^{\gamma(x,r_T)}\right], \quad (3)$$

where $\gamma(x, r_T)$ is the anomalous dimension of the target gluon distribution.

The main difference among the distinct phenomenological models comes from the behavior predicted for the anomalous dimension, which determines the transition from the nonlinear to the extended geometric scaling regime, as well as from the extended geometric scaling to the DGLAP regime.

In this contribution we restrict our analysis to the so called BUW model, which is able to describe the ep HERA data for the proton structure function and the hadron spectra measured in pp and dAu collisions at RHIC energies [11]. In the BUW model, the anomalous dimension is given by

$$\begin{aligned}\gamma(x, r_T) &= \gamma_s + \Delta\gamma(x, r_T) \\ &= \gamma_s + (1 - \gamma_s)(\omega^a - 1)\left[(\omega^a - 1) + b\right]^{-1}, \end{aligned} \quad (4)$$

where $\omega \equiv 1/(r_T Q_s(x))$, $\gamma_s = 0.628$ and the two free parameters $a = 2.82$ and $b = 168$ are fitted in such a way to describe the RHIC data on hadron production [11]. In comparison with other phenomenological parametrizations, in the BUW model, the behavior expected for the unintegrated gluon distribution in the large p_T limit (linear regime) is recovered: $\varphi(x_2, q_T) \propto 1/q_T^4$ at large q_T. In contrast, in Ref. [7] (called here the MTW model) the nonlinear effects were taken into account considering the model proposed long ago by Golec-Biernat and Wusthoff [12], where the forward dipole scattering amplitude is given by Eq. (3) with $\gamma = 1$. In this model the r_T integration in Eq. (2) can be carried out analytically and a simple expression for the unintegrated gluon distribution can be obtained:

$$\varphi(x_2, q_T) = 4\pi \frac{q_T^2}{Q_s^2(x_2)} \exp\left(-\frac{q_T^2}{Q_s^2(x_2)}\right). \quad (5)$$

Although this model satisfactorily describes the nonlinear regime (small-q_T), it clearly does not contain the expected behavior for large-q_T. Consequently, the resulting predictions are not valid at large values of the transverse momentum of the hadron.

3. Results

3.1. Net-baryon production

The net-baryon rapidity distribution is obtained integrating Eq. (1) in p_T between $p_{T_{min}} = 0$ and $p_{T_{max}} = \sqrt{s}\,e^{-y}$. The upper limit $p_{T_{max}}$ comes from the kinematical condition $x_F < 1$. Following Ref. [7] we assume that the nuclear valence quark distribution is given by $x\,q_v^A(x, Q^2) = N_{part}\,x\,q_v^p(x, Q^2)$, with N_{part} being the number of participants. The proton valence quark distribution is described by the MRST01-LO parametrization [13]. For the fragmentation function we use the KKP parametrization [14].

Substituting (5) by the unintegrated gluon distribution derived from the BUW model does not lead to significant changes in the final rapidity distributions, except at the lowest energies. In this energy region the rapidity distribution obtained with the BUW model does not show a dip at $y = 0$, in contradiction with the data (see references and discussions in [8]). This disagreement is an indication of the limitation of this approach at lower energies. On the other hand, at increasing energies, the MTW and BUW predictions for the net-baryon rapidity distributions in nucleus-nucleus collisions are essentially equivalent, even for central *PbPb* collisions at LHC energies. The same is true for proton-nucleus collisions [8].

As an illustration, in Fig.1 we present the results for nucleus-nucleus collisions at RHIC and LHC energies (the GBW curves below represent the predictions resulting from the substitution of the phenomenological model for the fragmentation function used in [7] by the KKP parametrization).

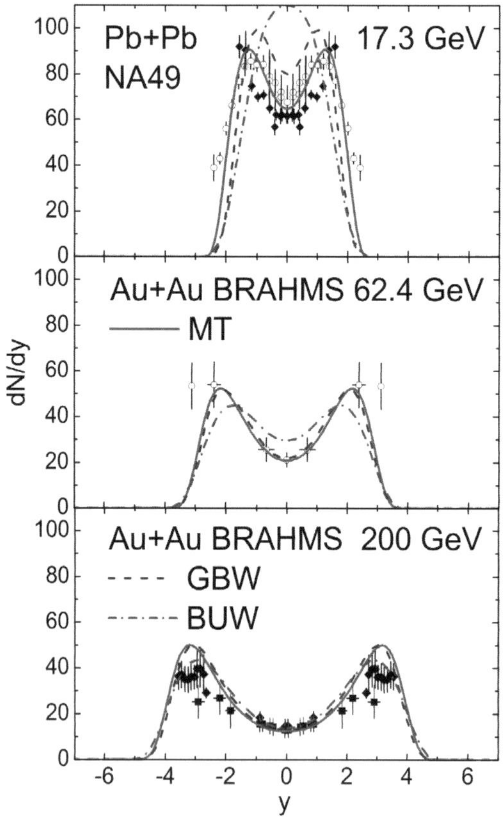

Figure 1. Net-baryon rapidity distributions in *PbPb* collisions at SPS energies and in *AuAu* collisions at RHIC energies. Data from [15, 16, 17, 18, 19, 20].

In Fig.2 we present our predictions for the net-baryon transverse momentum spectra in central *AuAu* collisions at RHIC energies. As in Ref. [7] we have assumed $N_{part} = 315$ and 357 for $\sqrt{s} = 62.4$ and 200 GeV, respectively. These plots show two striking features. First, we observe a very good agreement between data and the spectra obtained with Eq. (1) and the BUW dipole amplitude and, at the same time, a disagreement between data points and the spectra obtained with the GBW dipole amplitude, specially when $p_T > 1$ GeV. This happens because the GBW dipole amplitude has no DGLAP evolution and should not be able to reproduce data with large p_T. The BUW amplitude has the correct behavior at larger p_T and is able to describe the data in this region. Another interesting feature of these plots is the failure of the formalism at the largest rapidity and lowest energy. This may be an indication that here the baryons are not

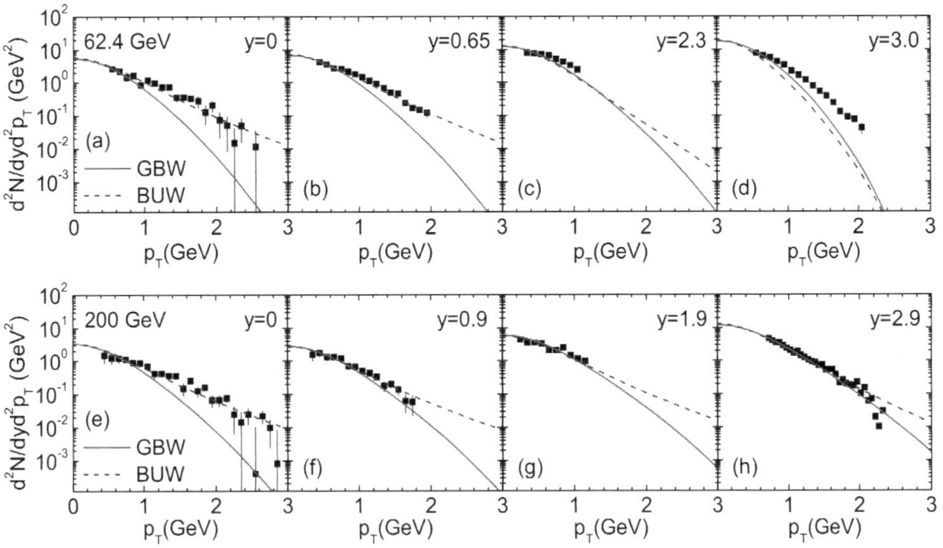

Figure 2. Net-baryon transverse momentum spectra in central $AuAu$ collisions at RHIC.

produced by independent quark fragmentation.

3.2. Leading particle spectra and Feynman scaling

Forward nucleon production is very important for cosmic ray physics, where highly energetic protons reach the top of the atmosphere and undergo successive high energy scatterings off light nuclei in the air. In each of these collisions, a projectile proton (the leading baryon) looses energy, creating showers of particles, and goes to the next scattering. The interpretation of cosmic data depends on the accurate knowledge of the leading baryon momentum spectrum and its energy dependence. The crucial question of practical importance is the existence or non-existence of the Feynman scaling, which says that x_F-spectra of secondaries are energy independent. In cosmic ray applications we are sensitive essentially to the large x_F region (the fragmentation region) and hence we can try to answer this question using the CGC formalism and the expressions derived in the preceding sections. An additional motivation for this calculation is the fact that, in the near future, Feynman scaling (or its violation) will be investigated experimentally at the LHC by the LHCf Collaboration [21, 22].

Changing variables from y to x_F and integrating (1) over p_T we obtain the x_F spectra of leading protons and pions in pp, pPb and $PbPb$ collisions at several energies, which are shown in Fig.3. In all panels we can clearly see a shift to smaller values of x_F, indicating a very slow softening of the leading particle spectrum. This approximate Feynman scaling is quite different from the behavior found in [6], where a much stronger softening of the LP spectra was observed. The difference between the two results can be traced back to difference between valence quark recombination and independent fragmentation. Whereas in the former energy loss is a complex phenomenon, which involves the whole baryon and changes more rapidly with the energy, in the latter a single valence quark interacts with the target and the LP spectrum changes only until the target reaches its high density limit. From this point on LP is approximately frozen.

The study of leading particle spectra will gain a new interest in view of the proposals of upgrades of the existing experimental facilities. In [23], the authors proposed an experiment

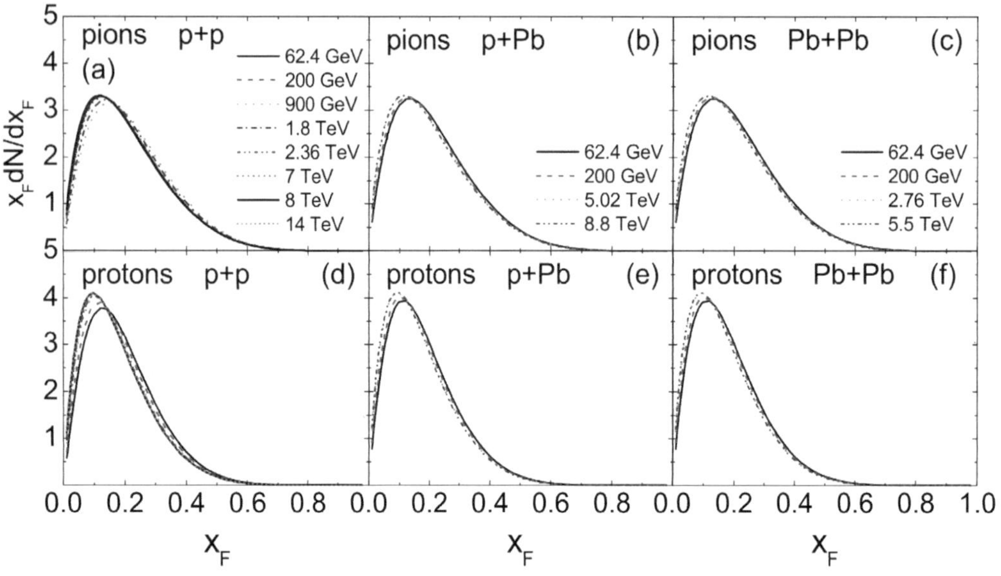

Figure 3. Energy dependence of the x_F distributions of pions and protons produced in pp, pPb and $PbPb$ colisions.

for the precise measurements of very forward particle production at RHIC. The proposal is to install a LHCf-like calorimeter in the ZDC installation slot at one of the RHIC interaction points. By installing a high-resolution electromagnetic calorimeter at this location it would be possible to measure the spectra of photons, neutrons and π^0 at pseudo rapidities above 6. The new measurements at 500 GeV p-p collisions would contribute to improve the hadronic interaction models used in the cosmic-ray air shower simulations. Using a similar kinematic coverage at RHIC to that of the measurements at LHC, it would be possible to test the Feynman scaling with a wide energy range and make the extrapolation of models into cosmic-ray energy more reliable.

4. Summary

In this work we have improved the CGC formalism of net-baryon production developed in [7] (a more detailed study of rapidity distributions, p_T and x_F spectra of forward protons and pions can be found in [8]). We obtain a good agreement with existing data and show predictions for the forthcoming LHC data.

Concerning forward proton production, our results suggest that at energies around $\sqrt{s} = 62.4$ GeV there is a transition from quark recombination to independent quark fragmentation. The independent fragmentation dynamics underpredicts the data at large rapidities and lower energies but starts to describe the data very well at higher energies. A solid conclusion about this change of mechanism still requires further theoretical and experimental work.

Finally, in the CGC formalism with independent valence quark fragmentation we observe an approximate Feynman scaling at very high energies.

Acknowledgments
This work was partially financed by the Brazilian funding agencies CAPES, CNPq, FAPESP and FAPERGS.

References

[1] D.S. Barton et al., Phys. Rev. D **27**, 2580 (1982); A.E. Brenner et al., Phys. Rev. D **26**, 1497 (1982); EHS/NA22 Collaboration, N.M. Agababyan et al., Z. Phys. C **75**, 229 (1996); N. Cartiglia, AIP Conf. Proc. **407**, 515 (1997).
[2] F. D. Aaron *et al.* [H1 Collaboration], Eur. Phys. J. C **68**, 381 (2010); J. Olsson [H1 Collaboration], PoS DIS **2014**, 156 (2014); V. Andreev *et al.* [H1 Collaboration], Eur. Phys. J. C **74**, 2915 (2014).
[3] A. Berera et al., Phys. Lett. B **403**, 1 (1997).
[4] F. O. Duraes, F. S. Navarra and G. Wilk, Braz. J. Phys. **35**, 3 (2005); Phys. Rev. D **58**, 094034 (1998); Phys. Rev. D **47**, 3049 (1993).
[5] A. Dumitru, L. Gerland and M. Strikman, Phys. Rev. Lett. **90**, 092301 (2003); [Phys. Rev. Lett. **91**, 259901 (2003)].
[6] F. O. Duraes, V. P. Goncalves, F. S. Navarra, A. L. V. R. dos Reis and G. Wilk, Braz. J. Phys. **37**, 122 (2007).
[7] Y. Mehtar-Tani and G. Wolschin, Phys. Rev. Lett. **102**, 182301 (2009); Phys. Rev. C **80**, 054905 (2009); Phys. Lett. B **688**, 174 (2010); Europhys. Lett. **94**, 62003 (2001).
[8] F. O. Durães, A. V. Giannini, F. S. Navarra and V. P. Gonçalves, Phys. Rev. C **89**, 035205 (2014).
[9] J. Jalilian-Marian, A. Kovner, L. McLerran and H. Weigert, Phys. Rev. D **55**, 5414 (1997); A. Kovner, J. G. Milhano and H. Weigert, Phys. Rev. D **62**, 114005 (2000); E. Ferreiro, E. Iancu, A. Leonidov, L. McLerran, Nucl. Phys. A **703**, 489 (2002).
[10] I. Balitsky, Nucl. Phys. B **463**, 99 (1996); Y. V. Kovchegov, Phys. Rev. D **60**, 034008 (1999); Phys. Rev. D **61**, 074018 (2000).
[11] D. Boer, A. Utermann and E. Wessels, Phys. Rev. D **77**, 054014 (2008); V. P. B. Gonçalves and M. L. L. da Silva, Nucl. Phys. A **906**, 28 (2013).
[12] K. Golec-Biernat and M. Wusthoff, Phys. Rev. D **59**, 014017 (1998); Phys. Rev. D **60**, 114023 (1999).
[13] A. D. Martin, W. J. Stirling and R. S. Thorne, Phys. Lett. B **636**, 259 (2006).
[14] B. A. Kniehl, G. Kramer and B. Pötter, Nucl. Phys. B **582**, 514 (2000).
[15] I. G. Bearden *et al.* [BRAHMS Collaboration], Phys. Rev. Lett. **93**, 102301 (2004).
[16] H. H. Dalsgaard, Int. J. Mod. Phys. E **16**, 1813 (2007).
[17] I. C. Arsene *et al.* [BRAHMS Collaboration], Phys. Lett. B **677**, 267 (2009).
[18] H. Appelshauser *et al.* [NA49 Collaboration], Phys. Rev. Lett. **82**, 2471 (1999).
[19] R. Debbe [BRAHMS Collaboration], J. Phys. G **35**, 104004 (2008).
[20] F. Videbaek and O. Hansen, Phys. Rev. C **52**, 2684 (1995); L. Ahle *et al.* (E802 Collaboration/Experiment E-866), Phys. Rev. C **60**, (1999) 064901; B.B. Back *et al.* (E917 Collaboration), Phys. Rev. Lett. **86**, 1970 (2001).
[21] T. Sako [LHCf Collaboration], arXiv:1010.0195 [hep-ex].
[22] O. Adriani, L. Bonechi, M. Bongi, G. Castellini, R. D'Alessandro, A. Faus, K. Fukatsu and M. Haguenauer *et al.*, Nucl. Phys. Proc. Suppl. **212**, 270 (2011).
[23] Y. Itow, H. Menjo, G. Mitsuka, T. Sako, K. Kasahara, T. Suzuki, S. Torii and O. Adriani *et al.*, arXiv:1401.1004 [physics.ins-det]; Y. Itow, H. Menjo, T. Sako, N. Sakurai, K. Kasahara, T. Suzuki, S. Torii and O. Adriani *et al.*, arXiv:1409.4860 [physics.ins-det]; L. Bonechi, O. Adriani, E. Berti, M. Bongi, G. Castellini, R. D'Alessandro, M. Del Prete and M. Haguenauer *et al.*, EPJ Web Conf. **71**, 00019 (2014).

Parton-hadron matter at finite temperature and chemical potential

E L Bratkovskaya[1], V P Konchakovski[2], W Cassing[2]

[1] Institute for heoretical Physics, niversity of Frankfurt, Frankfurt, Germany
[2] Institute for heoretical Physics, niversity of Giessen, Giessen, Germany
[3] e as A& niverstity, e as, .S.A.

-mail Elena.Bratkovskaya@th.physik.uni-frankfurt.de

Abstract. he dynamics of partons and hadrons in ultra-relativistic nucleus-nucleus collisions is analy ed within the Parton- adron-String Dynamics P SD transport approach, which is based on a dynamical uasiparticle model for the partonic phase DQP including a dynamical hadroni ation scheme while reproducing lattice Q D results in thermodynamic e uilibrium for the e uation-of-state as well as transport coefficients like shear and bulk viscosities or the electric conductivity of the hot Q D medium. he P SD model reproduces a large variety of observables from SPS to energies, e.g. the uark-number scaling of elliptic flow, transverse mass and rapidity spectra of charged hadrons, dilepton spectra, open and hidden charm production, collective flow coefficients etc., which are associated with the observation of a strongly interacting QGP sQGP . he highlights of the latest results from energies are presented with a focus on the correlation between the average transverse momentum $< p_T >$ and the number of charged particles N_{ch} in p-p, p-Pb and Pb-Pb collisions at midrapidity as observed by the A I ollaboration.

1. Introduction

he dynamics of the early universe in terms of the ' ig ang' may be studied e perimentally by ultra-relativistic nucleus-nucleus collisions at Relativistic- eavy-Ion- ollider R I or arge- adron- ollider energies in terms of 'tiny bangs' in the laboratory. ith sufficiently strong parton interactions, the medium in the collision one can be e pected to achieve local e uilibrium after some initial delay and e hibit appro imately hydrodynamic ow 1, 2, 3 . In these collisions a new state of strongly interacting matter is created, being characteri ed by a very low shear viscosity η to entropy density s ratio, η/s, close to a nearly perfect uid 4, . attice D l D calculations , indicate that a crossover region between hadron and uark-gluon matter should have been reached in these e periments.

In case of nuleus-nucleus e.g. Pb Pb collisions the ideal or viscous hydro calculations the initial conditions – at some nite starting time of the order of .3 to . fm/c – have to be evaluated either in terms of the standard Glauber model or other initial state scenarios like the IP-glasma model , or the G approach 1 , 11, 12, 13 , respectively. Alternative initial state scenarios are a coherent or incoherent superposition of 'strings' 14 or ust an incoherent superposition of hard nucleon-nucleon collisions as in the wounded nucleon model N 1 . Di erences between the di erent initial state assumptions and dynamical evolutions thus have to be e pected. he applicability of ideal or viscous hydrodynamic models to proton-nucleus reactions for low multiplicity events, however, is very much debated. his also holds for hybrid models 1 , 1 as long as they employ a hydro phase. o our knowledge only microscopic transport approaches 1 , 1 allow to bridge the gap from p-p to p-A and A-A collisions in a uni ue way without introducing additional and less controlled parameters. e will here employ the microscopic Parton- adron-String Dynamics P SD transport approach 2 to analy e early reaction dynamics in the conte t of collective ow coefficients to shed some light on the sensitivity of these observables on di erent assumptions on the initial state dynamics.

he comple ity of heavy-ion collisions is reduced essentially in the case of proton-nucleus collisions owing to the e pected dominance of the initial state e ects. In 2 13 the rst A I measurement of the charged particle pseudorapidity density has been reported 21 for $|\eta| < 2$ in p-Pb collisions at a nucleon-nucleon center-of-mass energy $\sqrt{s_{NN}}$. 2 eV. he measurement has been compared to two sets of particle production models that describe similar measurements for other collision systems the saturation models employing coherence e ects 11, 12 and the two-component models combining perturbative D processes with soft interactions 22, 23 . A comparison of the model calculations with the data has shown that the results are model-dependent and predict the measured multiplicities only within 2 %. Accordingly, the restrictions imposed by the measured minimal bias pseudo-rapidity spectra $dN_c/d\eta$ are not sufficient to disentangle di erent models for the very early interaction stage of ultra-relativistic collisions. e will here employ the P SD approach to study p-Pb and Pb-Pb reactions on an event-by-event basis in order to obtain further information on correlations between di erent observables such as the correlation between the average transverse momentum $< p_T >$ and the number of charged particles N_{ch} at midrapidity as observed by the A I ollaboration.

2. PHSD @ LHC

e brie y recall the ingredients of the P SD which is a covariant dynamical approach for strongly interacting systems formulated on the basis of adano - aym e uations 24 or o -shell transport e uations in phase-space representation, respectively. In the adano - aym theory the eld uanta are described in terms of dressed propagators with comple selfenergies. hereas the real part of the selfenergies can be related to mean- eld potentials of orent scalar, vector or tensor type , the imaginary parts provide information about the lifetime and/or reaction rates of time-like particles 2 . Once the proper comple selfenergies of the degrees-of-freedom are known, the time evolution of the system is fully governed by o -shell transport e uations as described in Refs. 24, 2 . he P SD approach includes a dynamical hadroni ation scheme based on transition rates and incorporates elastic and inelastic interactions of hadrons in the nal e pansion stage as in the SD model 2 . his approach allows for a simple and transparent interpretation of lattice D results for thermodynamic uantities as well as correlators in the partonic stage and leads to e ective strongly interacting partonic uasiparticles with broad spectral functions. For a review on o -shell transport theory we refer the reader to Ref. 2 . Actual P SD results and their comparison with e perimental observables for heavy-ion collisions from the lower super-proton-synchrotron SPS to R I energies can be found in Refs. 2 , 2 , 2 including electromagnetic probes such as e^+e^- or $\mu^+\mu^-$ pairs 2 , 3 .

2.1. p-p reactions at the LHC

o e tend the P SD model to higher energies than $\sqrt{s_{NN}}$ 2 GeV at R I , we have additionally implemented the PY IA .4 generator 31 for initial nucleon collisions at energies. For the subse uent lower energy collisions the standard P SD model 2 is applied including PY IA v . with J S v .3 for the production and fragmentation of ets 32 , i.e. for $\sqrt{s_{NN}} \leq$ GeV 32 . In this way all results from P SD up to top R I energies are regained and a proper e tension to energies is achieved. At $\sim \sqrt{s_{NN}}$ GeV both PY IA versions lead to very similar results. In PY IA .4 we use the Innsbruck pp tune

Journal of Physics: Conference Series **636** (2015) 012016 doi:10.1088/1742-6596/636/1/012016

3 which allows to describe reasonably the p-p collisions at $\sqrt{s_{NN}}$ eV in the framework of the P SD transport approach. he overall agreement with e perimental data for the distribution in the charged particle multiplicity N_{ch} a , the charged particle pseudorapidity distribution b , the transverse momentum p_T spectra c and the correlation of the average p_T with the number of charged particles d is satisfactory cf. Fig. 1 .

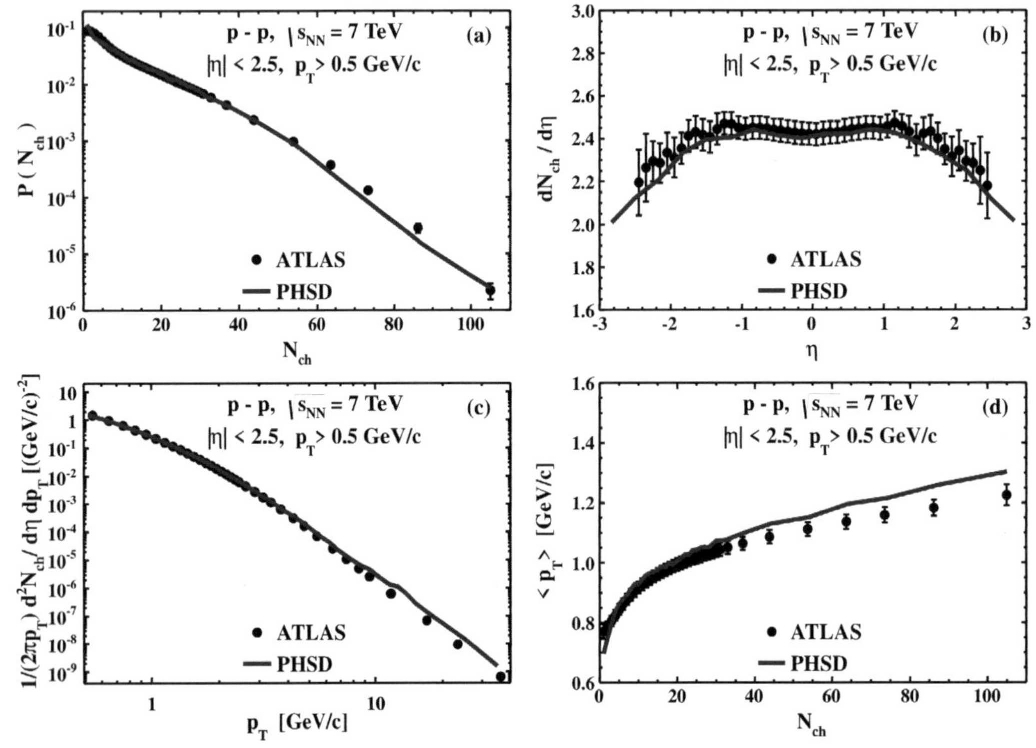

Figure 1. omparison of the P SD results including PY IA .4 with e perimental data from the A AS ollaboration 33 for p-p collisions at $\sqrt{s_{NN}}$ eV a N_{ch} distribution, b $dN_{ch}/d\eta$ distribution, c p_T-spectra and d average p_T vs. N_{ch}.

2.2. Properties of p-Pb collisions at the LHC

ith the elementary p-p collisions in the P SD being ad usted at energies via PY IA .4 using the Innsbruck pp tune 3 we proceed with observables and correlations from p-Pb collisions. In Fig. 2 lhs we present the probability distribution in the participant number N_{part} and the number of charged hadrons at midrapidity N_{ch} η as well as the ensemble average $< N_{ch} >$ vs. N_{part} rhs for p-Pb $\sqrt{s_{NN}}$. 2 eV . As is seen from Fig. 2 lhs , the number of charged particles at midrapidity correlates with the number of participants, N_{ch} η $\sim N_{part}$, however, with a large dispersion in both uantities. hen considering the ensemble average $< N_{ch}/d\eta >$ η vs. N_{part} we obtain the solid blue line within P SD while the dotted red line results from P SD when including additionally uctuations in the cross section P SD GG . For details we refer the reader to Ref. 34 . oth the standard Glauber and G results are presented and support the results of Ref. 1 . he two versions of the P SD model, with red dotted line and without cross section uctuations blue solid line , predict that the multiplicity dependence turns out to be close to the G result but substantially di ers from the wounded nucleon model N for larger N_{part}. hus, multiplicity distributions do not allow us to disentangle the di erent initial states under discussion from the G and a Glauber version as in P SD. he reason of such a multiplicity suppression relative to the N is the energy-momentum conservation in P SD which on average results in a decrease of particle multiplicity in subse uent scatterings as compared to the primary interaction.

he pseudorapidty distributions of charged particles from p-Pb minimum bias collisions at $\sqrt{s_{NN}}$. 2 eV are compared with the e perimental data 21 in Fig. 3. he data are displayed in the laboratory system which is shifted with respect to the nucleon-nucleon center-of-mass by y_{cm} − .4 . he results of two versions of the parton-hadron string dynamics model P SD and P SD-GG di er only for backward-emitted particles and both versions are rather close to the measured data and the G result open circles . Note that there are no modi cations or free parameters in the P SD e cept the e tensions by PY IA .4 which implies that p-p, p-A and A-A collisions are consistently described from low SPS to energies within ∼ 1 % .

he G predictions from Ref. 13 , performed earlier for the upcoming p-Pb run at the , are plotted in the same gure open circles . his result is based on the alitsky-ovchegov e uation 3 which is the large-N_c limit of non-linear renormali ation group e uations such as the -JI hierarchy 3 which is tested with respect to e p data. An astonishing result is that the G and P SD results almost coincide again. Note that this minimum-bias distribution corresponds to the mean charged particle multiplicity at the given value of pseudorapidity η. owever, event uctuations of $dN_{ch}/d\eta$ are very large as demonstrated in Fig. 3 rhs . hus, the study of minimum-bias $dN_{ch}/d\eta$ does not allow to disentangle the initial state concepts described within the P SD and G approaches.

et us, furthermore, consider pseudorapidity distributions for ed high-multiplicity events. Such distributions for di erent centrality bins have been measured by the A AS collaboration 3 for p-Pb collisions at $\sqrt{s_{NN}}$. 2 eV. perimentally the centrality was de ned according to selected bins in the transverse energy. e have de ned corresponding bins in N_{ch} keeping the same percentage of the number of selected events as in 3 the bin partition is shown in Fig. 4 lhs and the relative contribution of di erent centralities is given in the legend

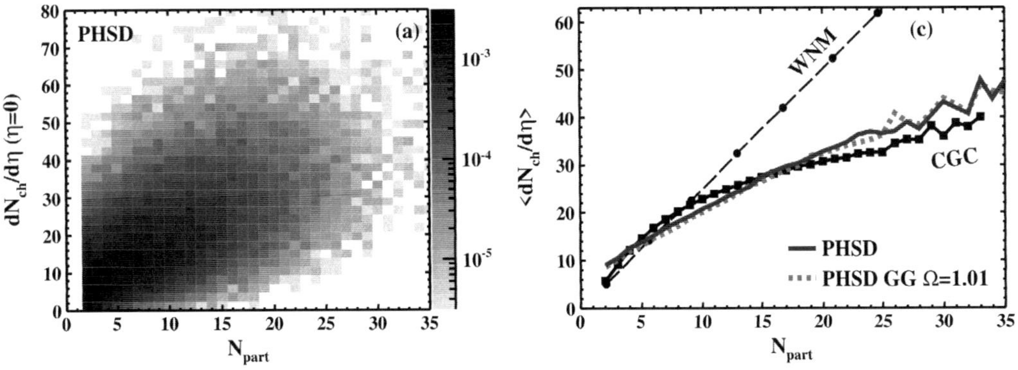

Figure 2. Probability distribution of the participant number and number of charged particles for Pb Pb at $\sqrt{s_{NN}}$. 2 eV at midrapidity lhs and its ensemle average in comparison to di erent models rhs . he wounded nucleon model N full dots and color glass condensate G calculations dotted line are taken from Ref 1 while simulations in the Glauber-Gribov appro imation full s uares and triangles stem from Ref. 3 . he P SD results are displayed in terms of the solid blue lines while the P SD results including uctuations in the cross section P SD GG are shown in terms of the dotted red lines.

Journal of Physics: Conference Series **636** (2015) 012016 doi:10.1088/1742-6596/636/1/012016

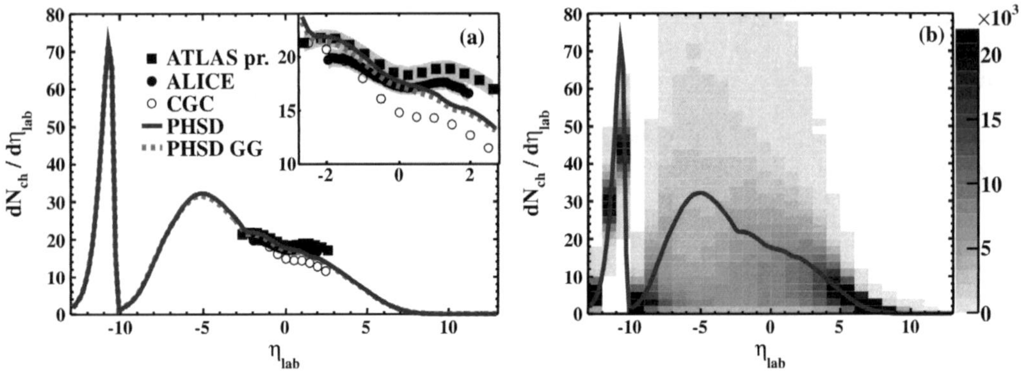

Figure 3. lhs Rapidity distribution of charged particles for minimum bias data from the A I 21 full dots and A AS 3 full s uares collaborations for p-Pb collisions at $\sqrt{s_{NN}}$. 2 eV in comparison to the P SD results solid blue line and the P SD GG results including uctuations in the cross section dotted red line . he G results open circles have been taken from Ref. 13 . he oomed results are displayed in the insertion. rhs vent-by-event uctuations of the rapidity distribution. he blue solid line shows the average charged particle pseudorapidity distribution.

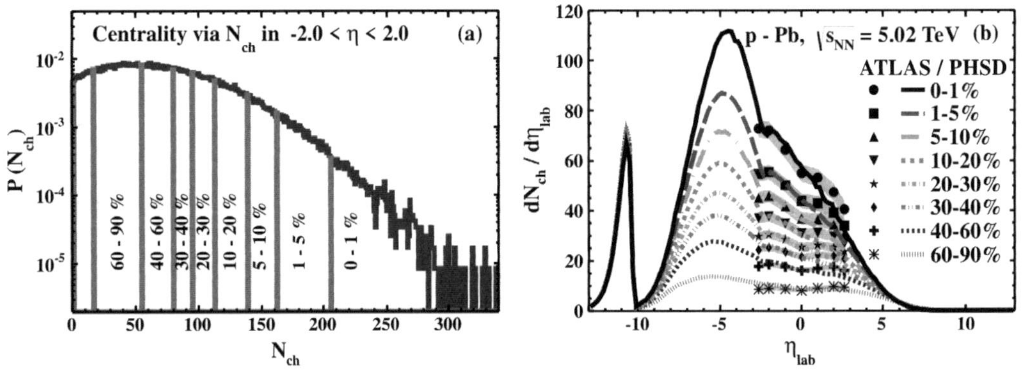

Figure 4. lhs entrality bins for p-Pb collisions at $\sqrt{s_{NN}}$. 2 eV selected according to the charged particle multiplicity in the rapidity interval $|\eta| < 2$. rhs omparison of the P SD calculated rapidity distributions with A AS data 3 for charged particles in di erent centrality bins. he shaded bands show the e perimental uncertainties.

in Fig. 4 rhs . In this gure the P SD results are based on 1 ^{6} simulated events.

As is seen from Fig. 4 rhs , the P SD model uite well reproduces the shape of the $dN_{ch}/d\eta$ distributions and its variation with centrality, in particular the increase with centrality of the forward-backward asymmetry between the directions of the proton-beam and Pb-target. For the most central events the P SD calculations very slightly overshoot this asymmetry, however, are in line with the data for the higher centralities within the e perimental uncertainties shaded areas in Fig. 4 rhs . e mention that the centrality sample of 4 - % with the ma imal number $N_{ch} \sim 2$ roughly corresponds to the minimum-bias distribution. For events of the highest multiplicity which amount to -1 % – corresponding to $\sim$ .1 ^{3} simulated events – the number of charged particles at the ma imum of the distribution is about . he agreement between calculations and data is not so bad taking into account the e perimental error bands and the fact that P SD has no free parameters once the p-p dynamics is ed by the PY IA tune .

2.3. Pb-Pb collisions at $\sqrt{s_{NN}}$=2.76 TeV

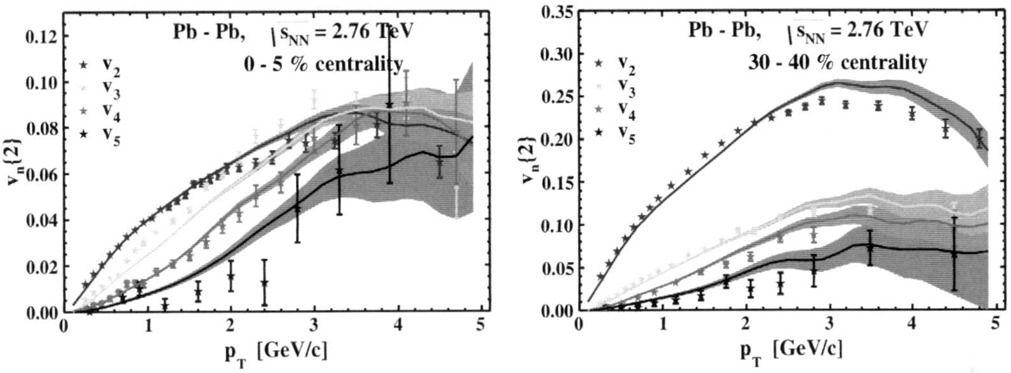

Figure 5. he ow coefficients v_2, v_3, v_4 and v_5 of all charged particles as a function of p_T for the centralities - % lhs and 3 -4 % rhs . he A I data have been taken from Ref. 3 .

Some general results for Pb-Pb collisions from the P SD approach have been reported in Ref. 42 . hereas the transverse charged single-particle spectra compare uite well with the e perimental observations at the 42 the uestion about the collective behavior of the system is of special interest. Anisotropic ow coefficients in both cases – i.e. e perimental data and P SD calculations – have been obtained from the two-particle cumulant method 43 in the central pseudorapidity window $|\eta| <$. and denoted as $v_n\{2\}$. In Fig. we compare the ow coefficients v_2, v_3, v_4 and v_5 of all charged particles from P SD as a function of p_T for the centralities - % lhs and 3 -4 % rhs in comparison to the A I data from Ref. 3 . he P SD results for v_2 p_T , v_3 p_T and v_4 p_T compare reasonably up to about 3. GeV/c whereas at higher transverse momenta the statistics is insufficient to draw robust conclusions. his also holds for the ow coefficient v_5 which still is in line with the data within error bars. It is uite remarkable that the collective behavior is reproduced not only for semi-central collisions rhs but also for - % central collisions lhs that are more sensitive to the initial uctuations 1 . hese tests indicate that the 'soft' physics at in central A-A reactions is very similar to the top R I energy regime although the invariant energy is higher by more than an order of magnitude. Furthermore, the P SD approach seems to work from lower SPS energies up to energies for p-p, p-A as well as A-A collisions, i.e. over a range of more than two orders in $\sqrt{s_{NN}}$.

e note in passing that a detailed study of the sensitivity of the initial state uctuations in space on the collective ow coefficients v_2, v_3 and v_4 turned out to be negative even for very central b Pb-Pb collisions at the cf. Ref. 42 which is essentially due to the fact that the global shape characteristics do not change much when increasing the local uctuations. hus our studies do not support the suggestion that the collective ow coefficients might be used to discriminate G from Glauber initial state conditions.

Journal of Physics: Conference Series **636** (2015) 012016 doi:10.1088/1742-6596/636/1/012016

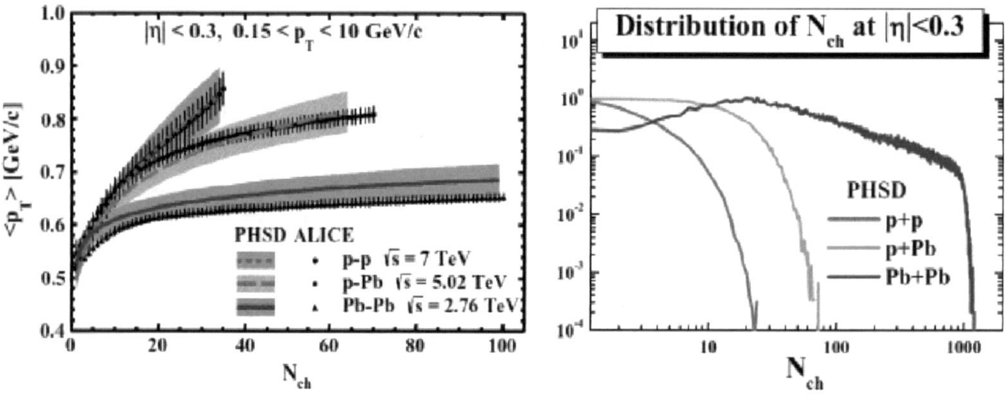

Figure 6. lhs Average p_T results for p-p, p-Pb and Pb-Pb collisions from the P SD transport approach 42 in comparison to the A I e perimental data from Ref. 44 . rhs he distribution in the number of charged particles for pseudo-rapidity $|\eta| <$ .3 from P SD for p-p, p-Pb and minimum bias Pb-Pb at the same energies as on the lhs.

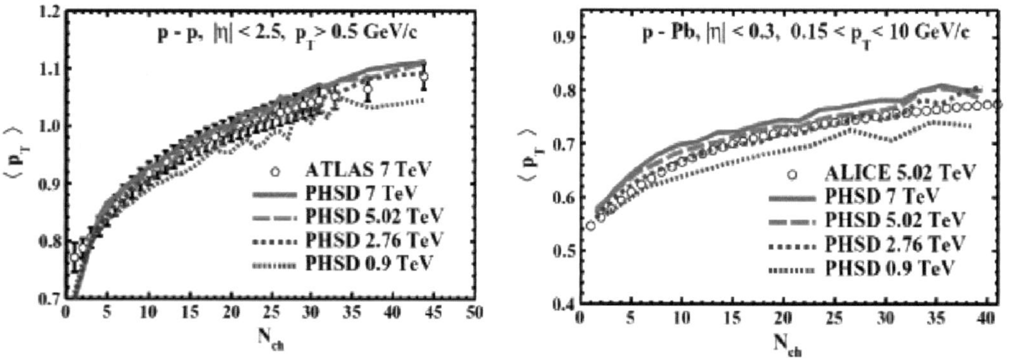

Figure 7. he correlation $< p_T >$ versus the number of charged hadrons at midrapidity for di erent bombarding energies for $p - p$ lhs and $p - Pb$ collisions rhs from . to eV. Note the di erent pseudo-rapidity and transverse momentum cuts!

2.4. Correlations between the transverse momentum and the number of charged particles at midrapidity

As a ne t step we will look for correlations and see how they evolve from p-p to p-Pb and Pb-Pb collisions. In Fig. lhs we compare the average $< p_T >$ as a function of charged particle multiplicity N_{ch} in p-p reactions at $\sqrt{s_{NN}}$ eV, p-Pb collisions at $\sqrt{s_{NN}}$. 2 eV and Pb-Pb collisions at $\sqrt{s_{NN}}$ 2. eV from the P SD to the e perimental data from Ref. 44 . Note that for low multiplicities $N_{ch} <$ the mean p_T is almost independent on the energy and the system see also Ref. 44 which in P SD can be traced back to the fact that for the acceptance $|\eta| \leq$ .3, .1 $\leq p_T \leq 1$ GeV/c only events with one or two binary collisions N_{bin} are selected for all systems. owever, with increasing number of charged particles N_{ch} the $< p_T >$ evolves uite di erently for p-p, p-Pb and Pb-Pb collisions which was hard to get in traditional approaches. Note, however, that the distribution in the charged particle multiplicity is very di erent for p-p, p-Pb and minimum bias Pb-Pb collisions rhs . his shows that only very peripheral Pb-Pb collisions may contribute to the correlation shown in Fig. lhs with a low number of binary collisions.

At rst sight one might attribute the very di erent correlations to the lower bombarding energies for p-Pb and Pb-Pb collision, however, this does not hold true since the correlation for p-p and p-Pb show practically the same low energy dependence as seen from the P SD calculations in Fig. .

o shed further light on the uestion we have a look at the transverse momentum of all charged particles for p-Pb at $\sqrt{s_{NN}}$. 2 eV from the P SD approach as a function of N_{ch} for the acceptance $|\eta| \leq$ .3, .1 $\leq p_T \leq 1$ GeV/c on an event-by-event basis Fig. lhs . ere the uctuations in p_T are large compared to the ensemble average $< p_T >$ orange line especially for low N_{ch}. Accordingly, e.g. a transverse momentum p_T . GeV/c does not correlate at all with N_{ch}! Furthermore, in these events for p-Pb the number of hard binary collisions N_{coll} varies from 1 up to numbers even above 1 Fig. rhs when e pecting at least one charged particle within the acceptance restricted to a very low window in pseudorapidity. For this event class the average number of hard collisions is $N_{coll} \approx 4$. e note in passing that patterns very similar to Fig. lhs are also found for Pb-Pb collisions at $\sqrt{s_{NN}}$ 2. eV.

In order to provide some global views on the reaction dynamics we display in Fig. the intensity in the events lhs and the ensemble average $< p_T >$ rhs as a function of N_{coll} and N_{ch} from P SD for $p - Pb$ collisions at $\sqrt{s_{NN}}$. 2 eV. hereas the average number of charged particles $< N_{ch} >$ scales linearly with the average number of hard collisions $< N_{coll} >$ as e pected this does not hold by far for the individual events! Furthermore, the average transverse momentum $< p_T >$ for ed N_{coll} and N_{ch} is roughly constant rhs and thus shows practically no correlations also with N_{coll}. Note that a ed N_{ch} or $< p_T >$ can be obtained by reactions with a varying number of binary collisions N_{coll}. ach of these binary reactions for larger N_{coll} then has a low N_{ch} and $< p_T >$, respectively. he ensemble averages nally lead to the average correlations shown in Fig. lhs .

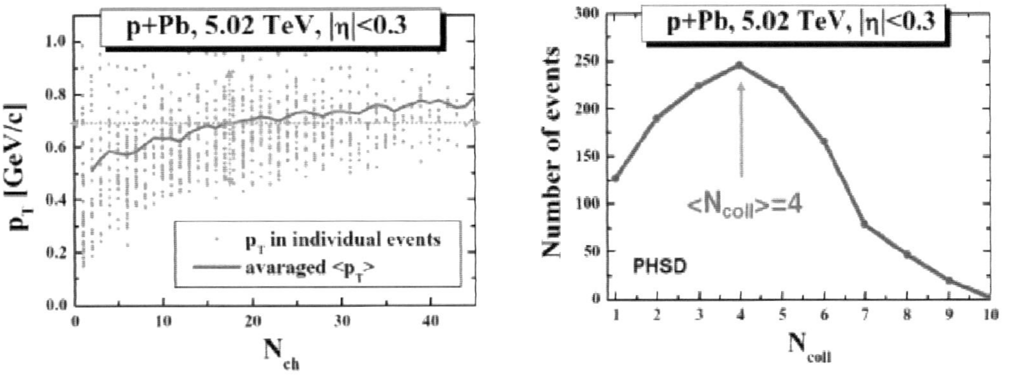

Figure 8. he correlation p_T versus the number of charged hadrons N_{ch} from P SD for $|\eta| <$ .3 on an event-by-event basis. he ensemble average $< p_T >$ is given by the orange solid line. rhs he number ofevents with at least one charged particle within the narrow acceptance as a function of the number of hard binary collisions N_{coll} for $p - Pb$ collisions at $\sqrt{s_{NN}}$. 2 eV.

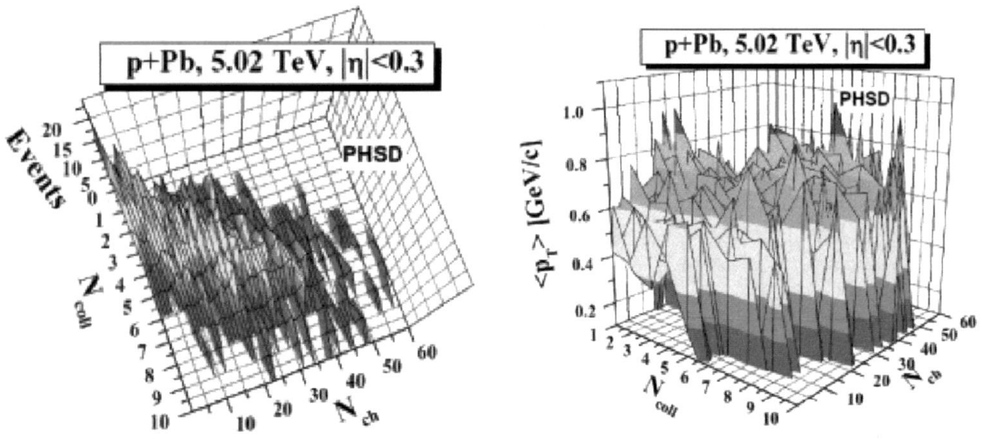

Figure 9. he intensity in the events lhs and the ensemble average $< p_T >$ rhs as a function of N_{coll} and N_{ch} from P SD for $p - Pb$ collisions at $\sqrt{s_{NN}}$. 2 eV within the acceptance $|\eta| \leq$ .3, .1 $\leq p_T \leq 1$ GeV/c .

3. Conclusions

In this study the parton-hadron-string dynamics P SD approach has been employed in the energy range for Pb-Pb collisions at $\sqrt{s_{NN}}$ 2. eV as well as p-Pb collisions at $\sqrt{s_{NN}}$. 2 eV. e nd that this approach works reasonably for both systems with respect to charged particle spectra as well as collective ow coefficients v_2, v_3, v_4 and v_5 for di erent centralities with a uality comparable to that achieved at R I energies before 2 , 2 , 2 , 2 , 4 , 41 . Our nding implies that the 'soft' physics in Pb-Pb collisions at the and Au-Au interactions at the top R I energies – despite a factor of $\sim$ 14 in $\sqrt{s_{NN}}$ – is very similar and in line with the dynamical uasiparticle model D P that de nes the parton properties for P SD in e uilibrium. his nding is common with earlier studies using viscous hydro approaches with varying initial conditions . In Pb-Pb collisions the P SD calculations have shown no sensitivity on the initial si e of spatial uctuations for the ow harmonics v_2 to v_4 which on one hand can be traced back to the low interaction rate in the initial none uilibrium stage in P SD $\sim$ .3 fm/c where e ects from di erent granularities are already washed out to some e tent. his is di erent from hydro calculations with varying granularity that instantly start to convert uctuations in coordinate space to collective modes in momentum space. On the other hand our method for changing the si e of initial uctuations keeps the event shape in coordinate space appro imately invariant cf. 42 which - in line with hydrodynamics - leads to very similar ow coefficients v_n. e mention that the low interaction rate in this very early phase in P SD is common with the G concept and thus does not allow to disentangle or determine the e ective degrees-of-freedom in this 'pre-hydro' phase.

Furthermore, we have e amined in more detail the correlations in the average transverse momentum $< p_T >$ versus the number of charged particles N_{ch} at midrapidity for p-p, p-Pb and minimum bias Pb-Pb events as measured by the A I ollaboration in Ref. 44 at di erent energies. he reasonable reproduction of the data within P SD could be traced back to the large uctuations in the number of binary collisions N_{coll} for a ed number of charged particles N_{ch} as well as huge uctuations in p_T for ed N_{ch} on an event-by-event basis . Furthermore, the individual average $< p_T >$ only weakly correlates with N_{coll} and the number of charged particles N_{ch} within the narrow acceptance $|\eta| <$ .3.

Acknowledgments
his work in part was supported by the O center I for FAIR as well as by F.

4. References

1 Ollitrault 1 2 *Phys. Rev.* D **46** 22
2 ein and olb P 2 2 *Nucl. Phys.* A **702** 2
3 Shuryak V 2 *Prog. Part. Nucl. Phys.* **62** 4
4 Shuryak V 2 *Nucl. Phys.* A **750** 4
Gyulassy and c erran 2 *Nucl. Phys.* A **750** 3
heng *et al.* 2 *Phys. Rev.* D **77** 14 11
Aoki *et al.* 2 *JHEP* **0906**
Schenke ribedy P and Venugopalan R 2 12 *Phys. Rev.* **C 86** 34
Gale eon S Schenke ribedy P and Venugopalan R 2 13 *Phys. Rev. Lett.* **110** 123 2
1 c erran 2 *Nucl. Phys.* A **787** 1
c erran 2 *Int. J. Mod. Phys.* A **21** 4
11 Dumitru A har eev D evin and Nara 2 12 *Phys. Rev.* **85** 44 2
12 ribedy P and Venugopalan R 2 12 *Phys. Lett.* **710** 12
13 Albacete Dumitru A Fu ii A and Nara 2 13 *Nucl. Phys.* A **897** 1
14 Andrs oscoso A and Pa ares 2 14 *Phys. Rev. C* **90** 4 2
1 dak A and Skokov V 2 13 *Phys. Rev. Lett.* **111** 1 23 1
1 Nonaka and ass S A 2 *Phys. Rev.* **75** 14 2
1 Qin G- Petersen ass S A and uller 2 1 *Phys. Rev.* **82** 4 3
1 Plumari S Puglisi A Scardina F and Greco V 2 12 *Phys. Rev.* **86** 4 2
1 u and Greiner 2 *Phys. Rev.* **71** 4 1
2 assing and ratkovskaya 2 *Nucl. Phys.* A **831** 21
ratkovskaya *et al.* 2 11 *Nucl. Phys.* A **856** 1 2
21 Abelev *et al.* A I ollaboration 2 13 *Phys. Rev. Lett.* **110** 323 1
22 arnafoldi G G arrette Gyulassy evai P and opor Pop V 2 12 *Phys. Rev.* **85** 24 3
23 u R Deng - and ang -N 2 12 *Phys. Rev.* **86** 1 1
24 uchem S assing and Greiner 2 4 *Nucl. Phys.* A **743** 2
2 assing 2 *E. Phys. J. ST* **168** 3
2 assing and ratkovskaya 1 *Phys. Rept.* **308**
assing ratkovskaya and uchem S 2 *Nucl. Phys.* A **674** 24
ratkovskaya and assing 1 *Nucl. Phys.* A **619** 413
Geiss assing and Greiner 1 *Nucl. Phys.* A **644** 1
2 oneev V D *et al.* 2 12 *Phys. Rev.* **85** 34 1
2 onchakovski V P ratkovskaya assing oneev V D and Voronyuk V 2 12 *Phys. Rev.* **85** 11 2
2 innyk O assing anninen ratkovskaya and o 2 12 *Phys. Rev.* **85** 24 1
innyk O ratkovskaya O venchuk V assing and o 2 12 *Phys. Rev.* **84** 4 1
3 innyk O *et al.* 2 13 *Phys. Rev.* **C 87** 14
31 S ostrand renna S and Skands P 2 *JHEP* **0605** 2
32 engtsson - and S ostrand 1 *Comp. Phys. Commun.* **46** 43
33 Aad G *et al.* A AS ollaboration 2 11 *New J. Phys.* **13** 3 33
34 onchakovski V P assing and oneev V D 2 14 *J. Phys.* G **41** 1 4
3 alitsky I 1 *Nucl. Phys.* **463**
ovchegov V 1 *Phys. Rev.* D **60** 34
3 Gelis F Iancu alilian- arian and Venugopalan R, *Ann. Rev. Nucl. Part. Science* **60** 4 3
3 A AS collaboration, A AS- ONF-2 13-
3 Abelev *et al.* A I ollaboration 2 14 *Phys. Lett.* **736** 1
3 Aamodt *et al.* A I ollaboration 2 11 *Phys. Rev. Lett.* **107** 323 1
4 innyk O assing and ratkovskaya 2 14 *Phys.Rev.* **89** 34
innyk O onchakovski V P assing and ratkovskaya 2 13 *Phys. Rev.* **88** 34 4
41 onchakovski V P *et al.* 2 12 *Phys. Rev.* **85** 44 22
42 onchakovski V P assing and oneev V D 2 1 *J. Phys. G* **42** 1
43 iland ic A Snellings R and Voloshin S 2 1 *Phys. Rev.* **83** 44 13
44 Abelev *et al.* A I ollaboration 2 13 *Phys. Lett.* **727** 3 1

Journal of Physics: Conference Series **636** (2015) 012017 doi:10.1088/1742-6596/636/1/012017

Boltzmann dynamics and temperature dependence of energy loss: Towards an understanding of the R_{AA} and v_2 puzzle for D-Mesons

F. Scardina [1,2], S. K. Das[1,2], S. Plumari[1,2] and V. Greco[1,2]

[1] Department of Physics and Astronomy, University of Catania, Via S. Sofia 64, I- 95125 Catania, Italy

[2] Laboratori Nazionali del Sud, INFN-LNS, Via S. Sofia 62, I-95123 Catania, Italy

E-mail: scardinaf@lns.infn.it

Abstract. The two key observables related to heavy quarks that have been measured in experiments are the nuclear suppression factor R_{AA} and the elliptic flow v_2. Reproducing these two observables simultaneously is a puzzle which have challenged all the existing models. We have studied two ingredients responsible to address a large part of such a puzzle: the temperature dependence of the energy loss and the full solution of the Boltzmann collision integral for the scattering between the heavy quarks and the particle of the bulk. We have considered four different models to evaluate the temperature dependence of drag and diffusion coefficients of the heavy quark. All these four different models are set to reproduce the same $R_{AA}(p_T)$ measured in experiments at RHIC and LHC energy. We have found that for the same $R_{AA}(p_T)$ one can generate 2-3 times more v_2 depending on the temperature dependence of the heavy quarks drag coefficient. Moreover comparing the Fokker-Planck and the Boltzmann approach we have found that the latter seems more efficient into reproducing the elliptic flow for the same R_{AA}. Even a larger difference between the two approaches emerges from the comparison of a more differential observable, the $c\bar{c}$ angular correlation.

1. Introduction

The ultra-Relativistic Heavy Ion Collisions (uRHIC) performed at the Relativistic Heavy Ion Collider (RHIC) and at the Large Hadron Collider (LHC) have given clear indications that a Quark Gluon Plasma has been created in laboratory. Beyond the light quark and gluons, which constitute the bulk, Heavy Quarks (HQ), charm and bottom, are created in the collisions. The interest in the study of HQ as probes to characterize the QGP [1, 2] is due to the fact that they are created essentially in the early stage of the collisions and since their masses are much larger than the temperature of the medium, they are decoupled from the bulk during the entire dynamical evolution of the system. Moreover their masses are larger than the typical momentum exchanged during the collisions with the light quark and gluons and thus they undergone to a Brownian motion which can be described by means of the Fokker-Planck equations. In such an approach the interaction is encoded in the drag and in the diffusion coefficients. The two key observables related to HQ that have been measured in experiments are the nuclear suppression factor R_{AA} and the elliptic flow v_2. However lately also a more differential observable, the $c\bar{c}$ angular correlation, has been proposed to characterize the QGP.

Several theoretical efforts have been made to reproduce the R_{AA} and the v_2 observed in experiments [3, 4, 5, 6, 7] within the Fokker Planck (FP) approach [8, 9, 10, 11, 12, 13, 14, 15, 16, 17, 18, 19, 20, 21] and the relativistic Boltzmann approach (BM)[22, 23, 24, 25, 26, 27, 28, 29]. However all the approaches show some difficulties to describe simultaneously R_{AA} and v_2. We have found that two ingredients assume a particular importance in reducing the differences between experimental data for R_{AA} and v_2 and theoretical calculations: the temperature dependence of the interaction and the use of full Boltzmann collision integral to study the time evolution of HQ momentum .

The proceeding is organized as follows. In section 2 we discuss the Fokker Planck approach which is used to describe the propagation of heavy quark through the QGP. In this section four different modelings to calculate the drag and diffusion coefficients, entailing different temperature dependence of the drag coefficients, will be presented. In section 4 and 5 we compare the FP approach and the BM approach analyzing three observables: R_{AA}, v_2 and $c\bar{c}$ angular correlation and finally section 6 contains the conclusions.

2. Fokker-Planck approach

The propagation of the Heavy quark is described by the Fokker Planck equations indicated here

$$\frac{\partial f}{\partial t} = \frac{\partial}{\partial p_i}\left[A_i(\mathbf{p})f + \frac{\partial}{\partial p_j}[B_{ij}(\mathbf{p})]\right] \tag{1}$$

where A_i and B_{ij} are the drag and the diffusion coefficients. To study the propagation of HQ this equation is replaced by an equivalent equation, the relativistic Langevin equation, which is more suited for numerical simulations

$$\begin{aligned} dx_i &= \frac{p_i}{E}dt, \\ dp_i &= -Ap_i dt + (\sqrt{2B_0}P_{ij}^{\perp} + \sqrt{2B_1}P_{ij}^{\parallel})\rho_j\sqrt{dt} \end{aligned} \tag{2}$$

where dx_i and dp_i are the coordinates and momenta changes in each time step dt; A is the drag force and B_0 and B_1 are respectively the longitudinal and the transverse diffusion coefficients; ρ_j is the stochastic Gaussian distributed variable; $P_{ij}^{\perp} = \delta_{ij} - p_i p_j/p^2$ and $P_{ij}^{\parallel} = p_i p_j/p^2$ are the transverse and longitudinal projector operators respectively. We employ the common assumption, $B_0 = B_1 = D$ [9, 10, 11, 13, 15, 16].

To solve the Langevin equation a background medium which describes the evolution of the bulk is necessary. Usually Hydrodynamical simulation are employed to describe the expansion and cooling down of the bulk. In this work instead the evolution of the bulk is provided by a 3+1D relativistic transport code tuned at fixed η/s [30, 31] which is able to reproduce the same results of hydrodynamical simulations. The transport code provides at each time step the density profile and the temperature profile of the bulk. The knowledge of the local density and of the local temperature at each time step are necessary to calculate the drag coefficients.

To generate the bulk profile we have performed simulations of $Au+Au$ collisions at $\sqrt{s} = 200$ AGeV for the minimum bias and $Pb+Pb$ at $\sqrt{s} = 2.76$ ATeV energy for centralities $0-20\%$ and $30-50\%$. In both cases the initial conditions in coordinate space are given by the Glauber model, while in the momentum space a Boltzmann-Juttner distribution function up to a transverse momentum $p_T = 2$ GeV has been considered. At larger momenta mini-jet distributions, as calculated within pQCD at NLO order, have been employed [32]. At RHIC energy, $Au + Au$ at $\sqrt{s} = 200$, the maximum initial temperature of the fireball at the center is $T_i = 340$ MeV and the initial time for the fireball simulations is $\tau_i = 0.6$ fm/c. At LHC instead the initial maximum temperature at the center of the fireball is $T_0 = 510$ MeV and the initial time for the simulations is $\tau_0 \sim 1/T_0 = 0.3$ fm/c.

HQ are initially distributed in coordinate space according to the number of binary collisions extracted from the Glauber model. The HQ distribution in momentum space is in accordance with the one in proton-proton that have been taken from [33]. In this present work four different modelings are used to calculate the drag coefficient, entailing different temperature dependence. The diffusion coefficient is instead calculated in accordance with the Einstein relation $D = \Gamma ET$. Our purpose is to investigate how for fixed R_{AA} the v_2 develops under different temperature dependence of the energy loss.

Model-I (pQCD) - We have considered elastic interaction among HQ and the bulk (light quarks and gluons). The scattering matrix related to these process $\mathcal{M}_{gHQ}$, $\mathcal{M}_{qHQ}$ and $\mathcal{M}_{\bar{q}HQ}$ in leading order are the well known Combridge matrix that includes s, t, u channel and their interferences terms [34]. The t-channel develops the well know infrared singularity that is regularized introducing a Debye screening mass $m_D = \sqrt{4\pi\alpha_s}T$ with a running coupling [35].
Model-II (AdS/CFT) - A drag force from the gauge/string duality [36] have been also considered in which the drag coefficient can be evaluated through the following equation

$$\Gamma_{conf} = C\frac{T^2_{QCD}}{M_c} \tag{3}$$

where $C = \frac{\pi\sqrt{\lambda}}{2\sqrt{3}} = 2.1\pm0.5$. Studies of dynamical evolution of heavy quarks within the Langevin dynamics using AdS/CFT can be found in Ref. [12, 19].
Model-III (QPM) - We have also evaluated the drag coefficient considering a bulk consisting of particles with a T-dependent quasi-particle masses, $m_q = 1/3g^2T^2$, $m_g = 3/4g^2T^2$. This model is able to successfully reproduce the thermodynamics of lQCD [37] (see also [38, 39]) by fitting the coupling $g(T)$. Such a fit leads to the following coupling [37]:

$$g^2(T) = \frac{48\pi^2}{[(11N_c - 2N_f)ln[\lambda(\frac{T}{T_c} - \frac{T_s}{T_c})]^2} \tag{4}$$

where λ=2.6 and T/T_s=0.57.
Model-IV ($\alpha_{QPM}(T), m_q = m_g = 0$) - We have also considered a case where the light quarks and gluons are massless but the coupling is from the QPM as indicated in Eq. 4. This last case has to be considered as an expedient to have a drag which decreases with the temperature.

For all the four cases considered the interaction has been rescaled to reproduce the R_{AA}

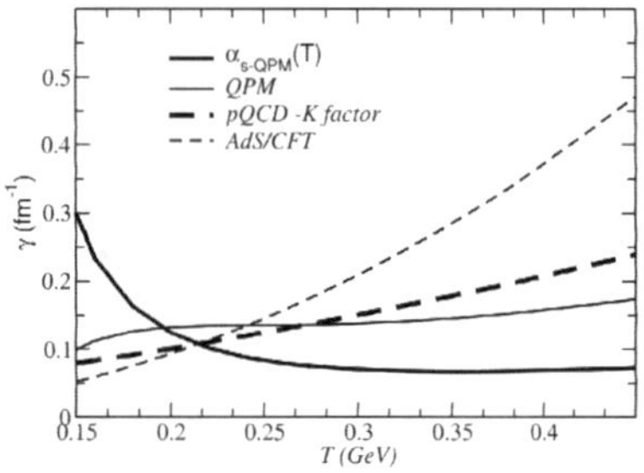

Figure 1. Variation of the drag coefficients with respect to temperature at $p = 5GeV$

observed in experiments. These rescaled drag coefficients have been shown in figure 1 as a function of temperature for the four different models. We observe that for all the modellings, with the exception of the *Model-IV* ($\alpha_{QPM}(T), m_q = m_g = 0$), the drag is an increasing function of the temperature. The dependence of the drag on T is proportional to T^2 for the AdS/CFT

while it is much milder for the QPM. The increasing of the drag with T in the *Model-IV* ($\alpha_{QPM}(T), m_q = m_g = 0$) is instead due to the increasing of the coupling at low temperatures. An increasing drag coefficient with the temperature implies that the interaction between the heavy quarks and the bulk is stronger in the latter stages of the fireball evolution.
Having the density and temperature profile of the bulk, provided by the Boltzmann code, we have simulated the Langevin dynamics for the four different models presented above. The Langevin equation gives as output the momentum distributions of HQ at the quark-Hadron transition temperature T_c. The momenta distributions are convoluted with the Peterson fragmentation functions of the heavy quark indicated in Eq. 5 in order to get the momentum distribution of D and B mesons.

$$f(z) \propto \frac{1}{[z[1-\frac{1}{z}-\frac{\epsilon_c}{1-z}]^2]} \tag{5}$$

where $\epsilon_c = 0.04$ for charm quarks and $\epsilon_c = 0.005$ for bottom quark. The ratio between the final distribution (f_f) and the initial distributions (f_i) of the D and B mesons is the nuclear modification factor $R_{AA} = f_i/f_i$ which is shown in figure 2 as a function of p_T for RHIC ($200AGeV$) while in figure 3 the elliptic flow ($v_2 = \langle((p_x^2 - p_y^2)/(p_x^2 + p_y^2))\rangle$) at the same energy as a function of p_T is depicted.

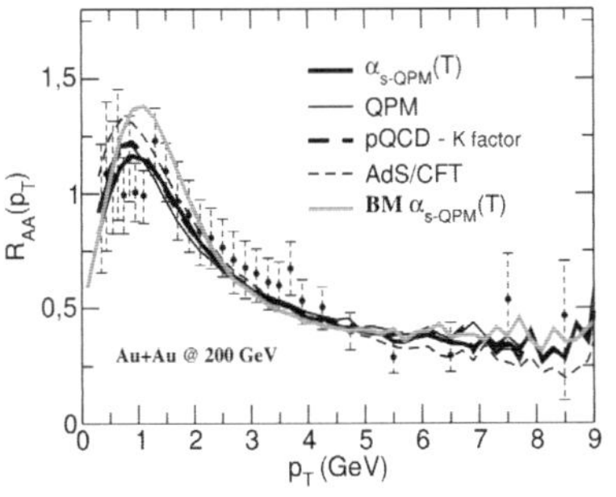

Figure 2. Comparison of the nuclear suppression factor as a function of p_T, obtained with the FP equation at RHIC for the four different T-dependences of the drag coefficient, with the experimental data. The orange line represents the result we obtain using the full Boltzmann equation for the $\alpha_{QPM}(T), m_q = m_g = 0$ case.

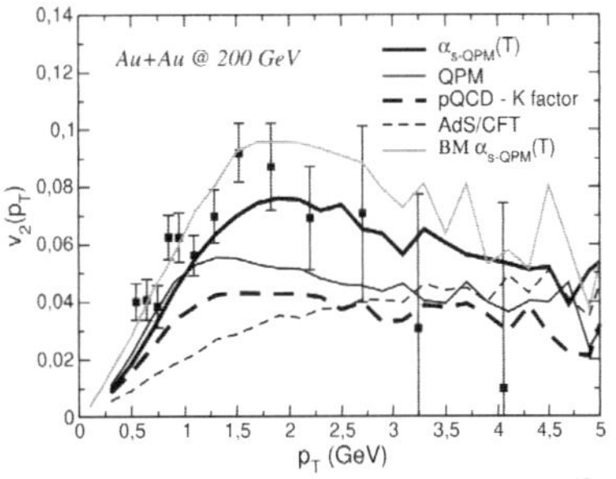

Figure 3. Comparison of the elliptic flow as a function of p_T, obtained with the FP equation at RHIC for the four different T-dependences of the drag coefficient, with the experimental data. The orange line represents the result we obtain using the full Boltzmann equation for the $\alpha_{QPM}(T), m_q = m_g = 0$ case.

In the figures 4 and 5 the R_{AA} and v_2 for LHC ($2.76ATeV$) are shown. As we already mentioned, we set to reproduce the same R_{AA} in all the cases by rescaling the drag coefficient. We observe that even if the R_{AA} is very similar for all the models the build-up of the v_2 is strongly related to the temperature dependence of the drag coefficient. In particular the larger is the interaction in the region of low temperature the larger is the elliptic flow. The same conclusions has been inferred also in the light flavor sector as shown in Refs. [40, 41, 42].

The reason of such a strong dependence of the elliptic flow on the temperature dependence of the drag coefficient is due to the fact that the elliptic flow is generated in the final stage of the evolution of the fireball when the temperature is lower. This is shown in the figures 6 and 7 where the R_{AA} and v_2 are shown at different times. We observe that the R_{AA} is generated

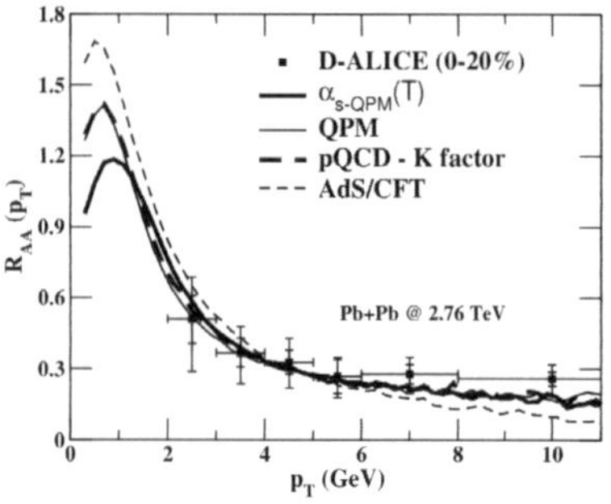

Figure 4. Comparison of the nuclear suppression factor as a function of p_T obtained within FP equation at LHC for the four different T-dependences of the drag coefficient with the experimental data.

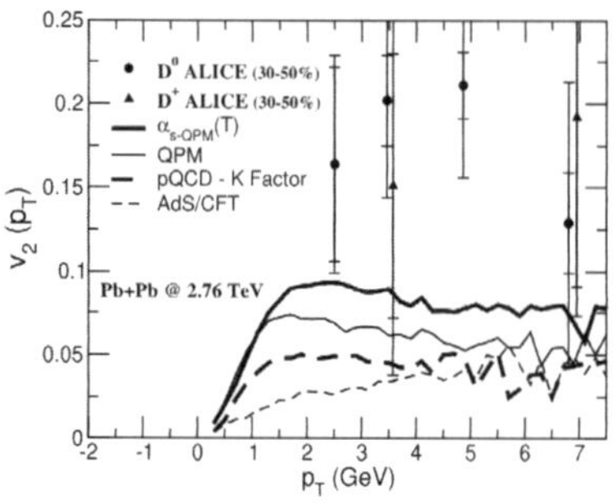

Figure 5. Comparison of the elliptic flow as a function of p_T obtained within FP equation at LHC for the four different T-dependences of the drag coefficient with the experimental data.

in the early stage of the QGP (large T) and is not sensitive to the final stage of the evolution, while the elliptic flow is generated later. This results refer to the pQCD case (*model I*) however the behavior is similar also for the other models.

This study suggests that in order to reduce the difference between the experimental data

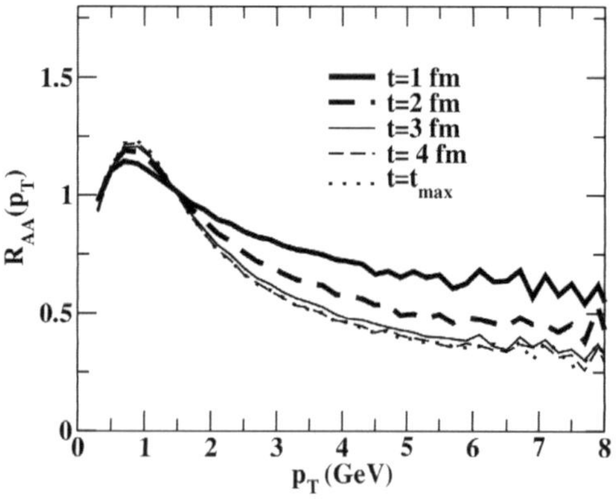

Figure 6. R_{AA} evaluated at different time at RHIC energies ($200AGeV$)

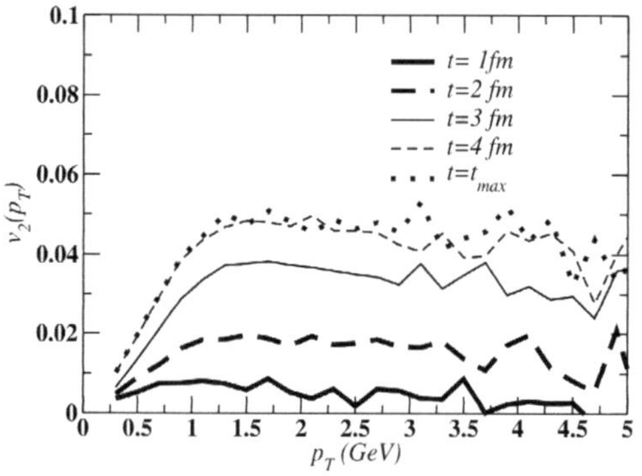

Figure 7. v_2 evaluated at different time at RHIC energies ($200AGeV$)

and the theoretical simulations the drag coefficient of the charm quark cannot increase with a large power of T. The case ($\alpha_{QPM}(T), m_q = m_g = 0$) is the one that can reproduce the experimental data but it is an extreme case used just to have a drag stronlgy decreasing with T. However also the QPM case can be considered quite close to the data considering that two other ingredients can have a substantial impact on the generation of the elliptic flow: the coalescence mechanism that would increase the v_2 for all the cases considered by about 20-25 % [11, 43] and the replacement of the FP equation, based on an expansion of the Boltzmann collision integral up to the second order, with the full collision integral that will be discussed in the next section.

Journal of Physics: Conference Series **636** (2015) 012017 doi:10.1088/1742-6596/636/1/012017

3. Boltzmann approach

The Boltzmann equation for the HQ distribution function is indicated here

$$p^{\mu}\partial_{\mu}f_{Q}(x,p) = \mathcal{C}[f_{Q}](x,p) \tag{6}$$

where $\mathcal{C}[f_Q](x,p)$ is the relativistic Boltzmann-like collision integral The collision integral can be written in a simplified form [1, 2] in the following way:

$$\mathcal{C}[f_Q](x,p) = \int d^3k[\omega(p+k,k)f_Q(x,p+k) - \omega(p,k)f_Q(x,p)] \tag{7}$$

where $\omega(p,k)$ expresses the collision rate of heavy quark per unit of momentum phase space. In this work we have not considered radiative processes therefore the heavy quarks interact with the medium by mean of two-body collisions regulated by the scattering matrix of the process $g,q+Q \rightarrow g,q+Q$, therefore defining the relative velocity between the two colliding particles as v_{rel} the transition rate can be written as:

$$\omega(p,k) = \int \frac{d^3q}{(2\pi)^3} f_{g,q}(x,p) v_{rel} \frac{d\sigma_{g,q+Q\rightarrow g,q+Q}}{d\Omega} \tag{8}$$

The Boltzmann equation is solved numerically dividing the space into a three-dimensional lattice and using the test particle method to sample the distributions functions. The collision integral is solved by means of a stochastic algorithm in which whether a collision happen or not is sampled stochastically comparing the collision probability $P_{22} = v_{rel}\sigma_{g,q+Q\rightarrow g,q+Q} \cdot \Delta t/\Delta x$ with a random number extracted between 0 and 1 [44, 45, 30, 46] . We use the Boltzmann equation to describe the propagation of the heavy quark as well as the evolution of the bulk as described in these reference [30, 47].

As emerge from Eq. 7 and from the expression of the collision probability an essential ingredient that has to be specified to study the propagation of the HQ using the BM equation is the cross section that can be calculated from the scattering matrix elements. From the same scattering matrix elements also the drag and diffusion coefficient, that are the keys ingredient of the FP equation, have been calculated. In such way we can make a comparison between the results we get using the BM and those we get using the FP approach having the same interaction.

The comparison between LV and BM approach has been thoroughly studied in these references [25, 26] where it is shown that for bottom quarks the FP is a very good approximation whereas for charm quark FP deviates significantly from the BM and such a deviation significantly depends on the the values of the Debye screening mass. We considered in references [25, 26] three values of m_D: 0.4 GeV, 0.83 GeV and 1.6 GeV. We have shown that the larger is m_D the larger is the difference between the FP approach and the Boltzmann approach. Here we have not considered a fixed value of the Debye screening mass but a value which depends on the temperature according to $m_D = gt$ where g have been evaluated through equation 4. The masses of light quark and gluons have been set equal to zero as it has been done using the Fokker-Planck approach with model IV described in section 2. In figures 2 and 3 the comparison for the RAA and v_2 at RHIC between the BM (orange thick line) and the FP (black thick line) are shown. We found that for the same R_{AA} we get a large v_2 using the BM with respect to the R_{AA} we get with the FP.

To summarize our results regarding the different values of elliptic flow that can be obtained using the different models of energy loss, we have introduced in Fig. 8 a new plot in which R_{AA} vs v_2 at a given momentum ($p_T = 1.3$ GeV) is depicted. This plot clearly shows how the building up of the v_2 can differ up to a factor 3 for the same R_{AA} depending on the temperature dependence of the energy loss and on the approach, BM or FP, used to describe the HQ propagation.

With the Boltzmann approach we get a value of the elliptic flow even larger with respect to the experimental data, however as already discussed in section 2, this case represent an extreme and not realistic case. Moreover to compare with experimental data also the coalescence hadronization mechanism has to be considered. However our results show a non-negligible impact of the approximation in the collision integral involved in the Fokker Plack equation on the relation between R_{AA} and v_2.

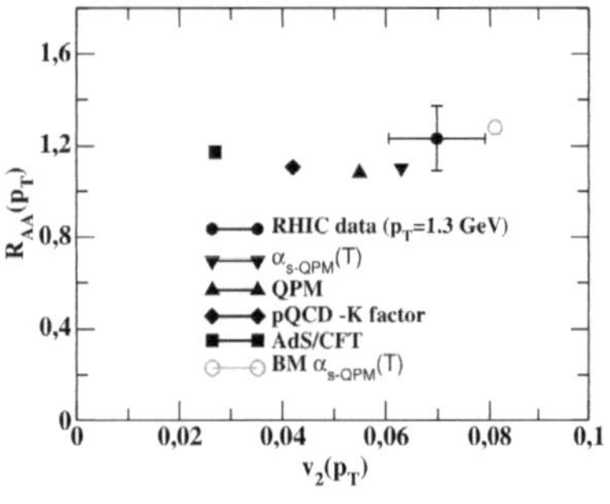

Figure 8. R_{AA} vs v_2 obtained with the FP for the four different T-dependences of the drag coefficient with the experimental data at RHIC energy at $p_T = 1.3GeV$. The open circle indicates the results obtained using the BM approach and the $\alpha_{QPM}(T), m_q = m_g = 0$ case

4. $c\bar{c}$ azimuthal correlations

To further investigate the differences between the FP and the BM approach we have also studied the $c\bar{c}$ correlations at both RHIC and LHC energies. We initialize the $c\bar{c}$ pairs initially in coordinate space accordingly to the binary collisions while in momenta space we have distributed them according to [33]. At the freeze out temperature we evaluate the azimuthal angles $\Delta\phi$ between c and $\bar{c}$ which we have initially distribute in the same point in coordinate space with $p_{T,c} = p_{T,\bar{c}}$, according to the LO initial production. We have performed the simulations in such way to have the same R_{AA} with the FP and the BM approach. We have carried out these simulations using $m_D = 1.6$ corresponding to a quite isotropic cross section which implies the maximum difference between the two approaches.

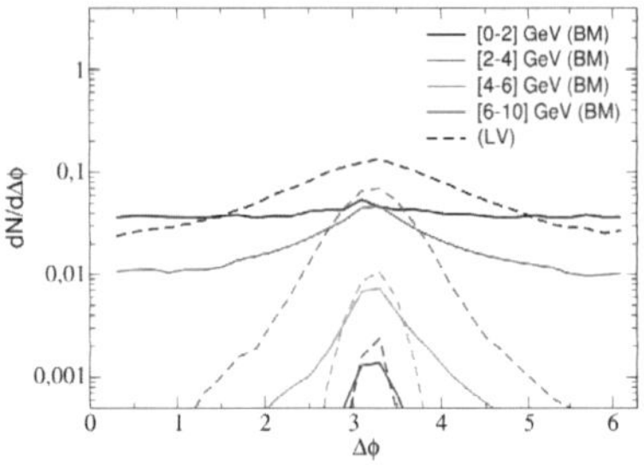

Figure 9. Comparison between $c\bar{c}$ azimuthal correlation at RHIC obtained using the BM (continue lines) and the FP (dashed lines) for different momentum range from $[0, 2]GeV$ (upper part of the figure) to $[6, 10]GeV$ (lower part)

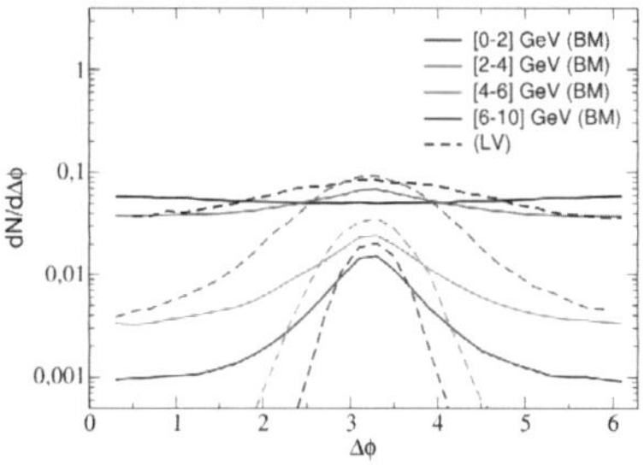

Figure 10. Comparison between $c\bar{c}$ azimuthal correlation at LHC obtained using the BM (continue lines) and the FP (dashed lines) for different momentum range from $[0, 2]GeV$ (upper part of the figure) to $[6, 10]GeV$ (lower part)

The final angular distribution are shown in Fig. 9, for RHIC and in Fig.10 for (LHC) considering four different p_T classes. The pairs are included in a class if both the charm and the anti charm are in the range of rapidity between -1 and 1 and if the momentum of the trigger particle (the one with larger transverse momentum) belongs to a given class. We notice that in all classes of momenta the initial correlations are broadened and the broadening is stronger using the Boltzmann approach. If one look at the lowest momentum class the initial correlation are almost completely washed out in the Boltzmann approach while in Fokker-Plank approach a residual correlation peak is still present. We observe substantial differences in the yield between the BM and the FP.
The reason for such a strong difference for the azimuthal correlations between BM and LV is due to larger spreading of momentum distribution in case of Boltzmann equation in comparison with the Langevin dynamics discussed in Refs [25, 26]. In practice the evolution of a single charm does not appear to be a simply shift in momentum with a Gaussian fluctuations around it. Moreover at LHC we observe in the BM approach an enhancement of the azimuthal correlation in the region of $\Delta\phi = 0$ that is due to the radial flow which push the $c\bar{c}$ in the same direction toward smaller opening angles as pointed out in this Refs [48, 49].

5. Conclusions

We have found that the temperature dependence of the interaction (drag coefficient) is an essential ingredient for the simultaneous reproduction of the nuclear suppression factor, R_{AA}, and elliptic flow, v_2 which is a current challenge almost for all the existing model. We find that typical T-dependence of the drag coefficients can lead to difference in v_2 by 2-3 times even if leading to the same R_{AA}. Our study suggests that the correct temperature dependence of the drag coefficient may not be larger power of T (as in pQCD or AdS/CFT) rather a lower power of T or may be constant in T.
Moreover we have studied the difference in the building up of the elliptic flow between FP and Boltzmann for a fixed R_{AA} and we observe that BM generates a larger elliptic flow [25, 26] than the FP. Finally we have also studied the $c\bar{c}$ correlations at both RHIC and LHC energies to further investigate the impact of the approximation involved in the FP equation. We have found a quite difference in the broadening of the initial correlations between the two approaches. We conclude that the more one look at differential observables (from R_{AA} to v_2 to $c\bar{c}$ azimuthal correlation) the larger is the difference in between FP and BM.

Acknowledgements

We acknowledge the support by the ERC StG under the QGPDyn Grant n. 259684

References

[1] Svetitsky B 1988 *Phys. Rev.* D **37**, 2484
[2] R.Rapp and H van Hees, R. C. Hwa, X. N. Wang (Ed.) Quark Gluon Plasma 4, 2010, *World Scientific*, **111**
[3] A. Adare *et al.* (PHENIX Collaboration), Phys. Rev. Lett. **98**, 172301 (2007).
[4] B. I. Abeleb (STAR Collaboration), Phys. Rev. Lett. **98**, 192301 (2007).
[5] A. Adare*et al.* (PHENIX Collaboration) A. Adare et al. (PHENIX Collaboration), Phys. Rev. C **84** , 044905 (2011)
[6] B. Abelev *et al.*,(ALICE Collaboration) JHEP **1209** (2012) 112
[7] B. Abelev *et al.*,(ALICE Collaboration) Phys. Rev.Lett. **111**,(2013) 102301.
[8] Mustafa M G, Pal D and Srivastava D K 1998 *Phys. Rev.* C **57**, 889
[9] Moore G D, Teaney D 2005 *Phys. Rev.* C **71** 064904
[10] Hees H van, Greco V and Rapp R 2006 *Phys. Rev.* C **73**, 034913
[11] Hees H van, Mannearelli M, Greco V and Rapp R 2008 *Phys. Rev. Lett.* **100** 192301
[12] Akamatsu Y, Hatsuda T, Hirano T 2009 *Phys. Rev.* C **79** 054907
[13] Das S K, Alam J, Mohanty 2009 *Phys. Rev.* C **80** 054916; 2010 *Phys. Rev.* C **82** 014908; Majumdar S, Bhattacharyya T, Alam J, Das S K 2012 *Phys. Rev.* C **84** 044901

[14] Alberico W M *et al.* 2011 *Eur. Phys. J.* C **71** 1666; Alberico W M *et al.* 2013 *Eur. Phys. J.* C **73** 2481
[15] S. Cao, G. Y. Qin and S. A. Bass, Phys. Rev. C **88** (2013) 4, 044907
[16] Lang T, Hees H, Steinheimer J and Bleicher M, arXiv:1208.1643 [hep-ph]
[17] M. He, R. J. Fries and R. Rapp, Phys. Rev. Lett. **110**, 112301 (2013)
[18] F. Riek and R. Rapp, Phys. Rev. C **82**, 035201 (2010)
[19] S. K. Das and A. Davody, Phys. Rev. C **89**, 054912 (2014)
[20] H-J Xu, X. Dong, L-J Ruan, Q. Wang, Z Xu, and Y Zhang, Phys.Rev. C **89** (2014) 024905
[21] S. K. Das, F. Scardina, S. Plumari and V. Greco, arXiv:1502.03757 [nucl-th].
[22] Gossiaux P B, Aichelin J 2008 *Phys. Rev.* C **78** 014904
[23] Uphoff J, Fochler O, Xu Z and Greiner C 2011 *Phys. Rev.* C **84** 024908
[24] J. Uphoff, O. Fochler, Z. Xu and C. Greiner, Phys. Lett. B **717** (2012) 430
[25] S. K. Das, F. Scardina, S. Plumari and V. Greco,Phys. Rev. C **90** 044901 (2014)
[26] F. Scardina,S. K. Das, S. Plumari and V. Greco, J.Phys.Conf.Ser. **535** (2014) 012019
[27] Younus M, Coleman-Smith C E,Bass S A and Srivastava D K, Phys.Rev. C **91** (2015) 2, 024912
[28] T. Song, H. Berrehrah, D. Cabrera, J. M. Torres-Rincon, L. Tolos, W. Cassing and E. Bratkovskaya, arXiv:1503.03039 [nucl-th].
[29] B. Zhang, L. -W. Chen and C. -M. Ko, Phys. Rev. C **72** (2005) 024906
[30] G. Ferini, M. Colonna, Di Toro M. and Greco V, 2009 *Phys. Lett.* **B670**,325; V. Greco ,M. Colonna , Di Toro M and Ferini G, 2009 *Progr. Part. Nucl. Phys.* **62**, 562
[31] S. Plumari,A. Puglisi, F. Scardina and V. Greco, Phys. Rev. C **86** (2012) 054902
[32] V. Greco, C. M. Ko and P. Levai, Phys. Rev. Lett. **90** (2003) 202302; V. Greco, C. M. Ko and P. Levai, Phys. Rev. C **68** (2003) 034904
[33] M. Cacciari, P. Nason and R. Vogt, Phys. Rev. Lett. **95** (2005) 122001; M. Cacciari, S. Frixione, N. Houdeau, M. L. Mangano, P. Nason and G. Ridolfi, JHEP **1210** (2012) 137
[34] B. L. Combridge, Nucl. Phys. B **151**, 429 (1979)
[35] O. Kaczmarek and F. Zantow, Phys. Rev. D, **71**, 114510(2005)
[36] J. M. Maldacena, Adv. Theor. Math. Phys. **2**, 231 (1998)
[37] S. Plumari, W. M. Alberico, V. Greco and C. Ratti, Phys. Rev. D, **84**, 094004 (2011)
[38] S. K. Das, V. Chandra, J. Alam, J. Phys. G **41** 015102 (2014)
[39] H. Berrehrah, E. Bratkovskaya, W. Cassing, P.B. Gossiaux, J. Aichelin, and M. Bleicher, Phys.Rev. C **89**, 054901 (2014)
[40] J. Liao and E. Shuryak, Phys. Rev. Lett. **102** (2009) 202302
[41] F. Scardina, M. Di Toro and V. Greco, Phys. Rev. C **82** (2010) 054901.
[42] X. Zhang and J. Liao, Phys. Rev. C **89**, no. 1, 014907 (2014); J. Xu, J. Liao and M. Gyulassy, arXiv:1411.3673
[43] R. J. Fries, V. Greco, P. Sorensen, Ann. Rev. Nucl. Part. Sci. **58**, 177 (2008)
[44] A. Lang *et al.*, Jour. of Comp. Phys. **106**, 391 (1993)
[45] Xu Z, Greiner C 2005 *Phys. Rev.* C **71**, 064901
[46] F. Scardina, D. Perricone, S. Plumari, M. Ruggieri and V. Greco Phys. Rev. C **90** (2014) 054904
[47] M. Ruggieri, F. Scardina, S. Plumari and V. Greco, Phys. Rev. C **89** (2014) 054914
[48] X. Zhu, N. Xu, and P. Zhuang Phys. Rev. Lett. **100** 152301 (2008)
[49] M.Nahrgang, J. Aichelin, P.B. Gossiaux and Klaus Werner,Phys. Rev. C **90** 024907 (2014)

Photon radiation from hot and dense matter of quark gluon plasma

Y Kumar[1], A Goel[2] and P Jain[2]

[1]Department of Physics, Sri Aurobindo ollege, niversity of Delhi, New Delhi, India
[2]Amity Institute of Nuclear Science and echnology, Amity niversity, Noida, India

-mail yogesh.du81@gmail.com

Abstract. e work on photon radiation from hot and dense matter of uark gluon plasma QGP . he total photon spectra of leading order O process is produced in relativistic massive collisions of two nuclei using various value of uark phenomenological flow parameter dependent on uark chemical potential in uark mass. he total photon measurement is strongly increasing function of uark flow parameter ranging $2\gamma_g \leq \gamma_q \leq \ \gamma_g$ and also nd appreciable enhancement at the Relativistic eavy Ion ollider R I at N and the arge adron ollider at RN as a result of high temperature and chemical potential using uark phenomenological flow parameter. he results obtained are presented on the study of photon production of leading order process. hus, the study of photon radiation from hot and dense matter provides a valuable information in the formation and evolution of QGP and in the subse uent investigation of its properties. he results are compared with other theoretical work.

1. Introduction

Since long, the study of massive nuclear collisions at R I and have become an important and uni ue situation to study the GP. he e periments in heavy ion collision e plore a new form of uantum chromodynamics matter where the net baryon density is non-vanishing with unusual properties in the early stage called the uark-gluon plasma GP 1, 2 . Now, this eld has entered into a new era of measurements, it is therefore important to consider the non vanishing chemical potential in a hot and dense nuclear matter of GP. In order to deal with hot and dense matter of GP, a precise study of heavy ion phenomenology is re uired 3 . Although there are many phenomenological models used to describe the properties of GP, a perfect evidence is still not con rmed.

he e periments have searched for a variety of signatures of a uark-gluon plasma. any indirect signatures have been suggested. Among all, photon radiation is one of the important phenomenological tool to study hot and dense matter of uark gluon plasma. hese photons are emitted throughout the evolution of a heavy-ion collision 4, . Since these valuable probes su er little interaction with the GP or hadronic matter, they carry information deep inside the hot and dense GP. hese particles do not e perience the strong force, su er little nal state interactions and having large mean free path in strongly interacting medium. his makes the photons an ideal probe of the GP − 1 .

herefore photons provide direct information via electromagnetic interaction from the collision one of plasma and escape to the detector undistorted through the hot and dense medium ,1 . he theoretical calculations nd a dominant contribution of the photons produced in the GP to the direct photon spectrum 2 – 22 . Apart from these advantages, these probes have some difficulties due to massive pollution from di erent sources.

In past decade, the works on photon radiation have been studied by ignoring the uark chemical potential 23 – 2 . After the con rmation of chemical potential at R I energies 2 ,3 , the thermodynamic e uilibrium is a function of both the temperature and chemical potential that in turn lead to the modi cation of photon production. Physicists calculated the photon rate of a GP at nite chemical potential 31, 32 . For more review on photon production in the conte t of nite uark chemical potential see Ref. 33 – 42 . Also iro, Strickland and their co-workers 43, 44 studied the photon emission in a chemically e uilibrating baryon free GP system.

any observables have been proposed to probe the GP. ere we e tend our work by considering the O processes for photon radiation. In this phenomenological model, we use nite uark mass that generated due to the interaction among uarks and/or gluons 23, 4 , 4 . At non vanishing chemical potential, this uark mass depends on temperature as well as chemical potential using ow parameters and used to avoid infrared divergences. he photon emission rate is observed using various initial condition taken from R I and energies. he uark mass is suitably modi ed using Ref. 23, 4

$$m_q^2 \; T, \mu_q \quad \gamma_q \left(1 \quad \frac{\mu_q^2}{\pi^2 T^2}\right) g^2 \; k \; T^2. \qquad 1$$

ere, $g^2 \; k \quad 4\pi\alpha_s$ with D strong coupling constant α_s de ned as;

$$\alpha_s \quad \frac{4}{33 - 2N_f \;\; \ln 1 \quad \frac{k^2}{\Lambda^2}}. \qquad 2$$

In the above e pression, k is de ned as

$$k \quad \left[\frac{\gamma N^{\frac{1}{3}} T^2 \Lambda^2}{2}\right]^{\frac{1}{4}}. \qquad 3$$

he parameter N is given as, $N \quad \frac{16\pi}{[33-2N_f]}$.

he earlier value γ_q is modi ed as $\gamma_q \; 1 \quad \mu_q^2/\pi^2 T^2$ for the good appro iamtion of temperature 4 . On the other side, we put gluon parameter remain same i.e. γ_g' as γ_g. he value of γ_g is ed as 1/3. he parametri ation factor $\gamma^2 \quad 2 \frac{1}{\gamma_q^2} \quad \frac{1}{\gamma_g^2}$. e the range of γ_q as $\gamma_q \quad 2\gamma_g$ to γ_g and $\gamma_g \quad 1/3$ in our calculation 4 . is the temperature and $g \; k$ is rst order D running coupling constant 23, 4 . sing modi ed uark mass, we study the photon radiation of leading order process in relativistic massive collisions of two nuclei using various value of uark phenomenological ow parameter dependent on uark chemical potential with avor $N_f \quad 2$.

he total photon spectra are calculated by integrating the production rate over the four-volume of the collision while taking into account the dynamics of the matter. e plot total photon production with low and intermediate transverse momentum region $k_T \leq$ GeV at temperature T . GeV and T . 3 GeV using various value of uark ow parameter dependent on uark chemical potential at R I and . It is uite certain that photons in the relevant range of transverse momentum provide valuable information in the hot and dense matter of GP. hus, the present work is interesting and important to account for these e ects in the measurements of photon radiation of O process at the hot phase of temperatures.

hus, we organi e the paper as In section 2, we present the photon radiation from hot and dense matter of GP. In last section 3, we give the results and conclusion.

2. Photon radiation from hot and dense matter of QGP

he study of photons have now become an e citing eld in high energy heavy ion collision. any theoretical and e perimental physicists all over the world are working in this eld to understand the best signatures of the GP formation and measuring them. he signi cant work has been done by many authors and lots of reasearch is on the way to study the hot and dense matter of GP. Since photons have been proposed as a promising signature of the GP in relativistic heavy-ion collision, our interest lies within hot and dense nuclear matter in heavy ion collision. he complete calculation of photon radiation from a partonic medium to order of α_s has been considered in the Ref. 24, 2 , 4 – . he relevant O processes include di erent channels like uark-anti uark annihilation, D compton scattering, remsstrahlung brems and uark-anti uark annihilation with scattering aws . For the present calculation, the photon rate of momentum k is given by the e pression 4 ,

$$\frac{dN}{d^4xd^3\mathbf{k}} \quad . \quad 4A\ k \left[ln\left(\frac{T}{m_q\ T, \mu_q}\right) \quad \frac{1}{2} ln\left(\frac{2k}{T}\right) \quad C_T\left(\frac{k}{T}\right)\right] \qquad 4$$

he leading-log coefficient $A\ k$ is given as

$$A\ k \quad . \ 2\ \frac{m_q^2\ T, \mu_q}{k} f_D\ k$$

ere $f_D\ k$ is the fermi distribution function. he dependence on the speci c photon production process is written in the term $C_T\ \frac{k}{T}$,

$$C_T\left(\frac{k}{T}\right) \quad C_{2\leftrightarrow 2}\left(\frac{k}{T}\right) \quad C_{brems}\left(\frac{k}{T}\right) \quad C_{aws}\left(\frac{k}{T}\right).$$

he non-logarithmic two-to-two contribution $C_{2\leftrightarrow 2}$ for general k/T and the rate of photon production by bremsstrahlung and annihilation with scattering are computed. his re uires solving a non-trivial integral e uation to determine this rate. he thermal corrections to the dispersion relations for incoming or outgoing particles can no longer be neglected in these $2 \leftrightarrow 2$ processes. he $C_T\ \frac{k}{T}$ is the non-trivial function that can only be solved numerically. All results for $C_T\ \frac{k}{T}$ are evaluated by Ref. 4 , .

e compute above photon rates of complete leading order over the space time volume of the reaction according to orken's model at temperature T . GeV and T . 3 GeV with the various value of uark ow parameter dependent on chemical potential for avor 2. he volume element $d^4x \quad d^2x_T dy \tau d\tau$, where τ is the evolution time of the system and y is the rapidity. y considering the central collisions, the integration over transverse coordinates yields a factor of $d^2x_T \quad \pi R_A^2$. e nally obtain the total photon spectra at hot temperature using Ref. 23, 42, 4 , . It is e pressed as

$$\frac{dN}{d^2k_T dy} \quad \pi R_A^2 \int_{\tau_0}^{\tau_f} \tau d\tau \int_{-y_{nuc}}^{y_{nuc}} dy \left(k \frac{dN}{d^3kd^4x}\right).$$

here τ is time evolution determined with the temperature from initial to nal state with corresponding rapidity value at R I and energy. he parameter k_T is the photon transverse momentum. hus, with the values of rapidity, k_T and other various set of initial condition, we obtain the total photon spectrum of O processes at R I and with uark avor $N_f \quad 2$.

3. Results and Conclusion

he theoretical knowledge of the production rates is an essential ingredient for the phenomenological description of photon signatures. In this contribution, we perform the calculation of O processes for photon radiation from hot and dense matter of GP which consist D ompton Scattering, uark-anti uark annihilation, bremsstrahlung and uark-anti uark annihilation with scattering using various initial condition at R I and with the e ect of phenomenological ow parameters such as γ'_q and γ_g dependent on uark chemical potential in uark mass.

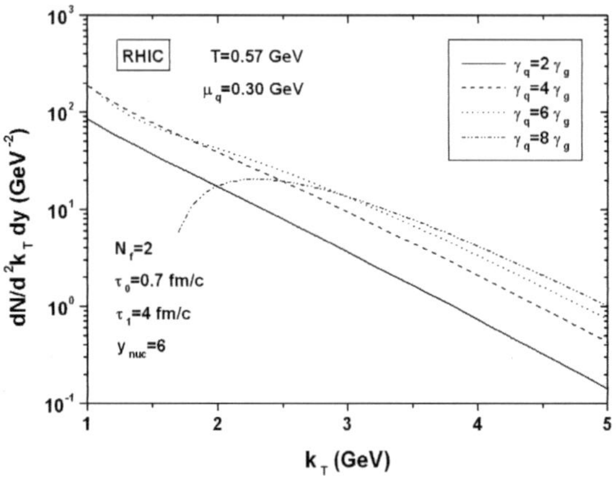

Figure 1. Photon Spectra at temperature T . GeV using various value of uark ow parameter dependent on uark chemical potential at R I .

he calculation of O processes for photon radiation is e tended with the suitable range of phenomenological parameter from γ_q $2\gamma_g$ to γ_g that dependent on nite value of uark chemical potential with temperature T . GeV and T . 3 GeV for avor N_f 2. e study the total photon spectrum over the space-time evolution of GP.

In Figure 1 , we plot total photon spectra for various value of phenomenological parameter ranging from γ_q $2\gamma_g$ to γ_g dependent on uark chemical potential μ_q .3 GeV at hot temperature T . GeV for avor N_f 2. e found that the total photon rate strongly increases with increases the uark phenomenological parameter at uark chemical potential μ_q .3 GeV and at hot temperature T . GeV. Also there is uniform fall in total emission rate as a function of transverse momentum k_T for all values of phenomenological parameter.

he strength of the photon radiation enhances due to the baryon rich plasma. he larger number of anticipating uarks brings more interactions, resulting enhancement in the production rate. It is also observed that the photons yield suppress more for the higher value of ow parameter and at γ_q γ_g, suppresion is more as comparison to the lower values. At closer inspection, the production rate is low for the value of ow parameter γ_q γ_g up to transverse momentum k_T 2 GeV and for γ_q γ_g up to k_T 3.2 GeV and afterwards photon rate increases.

he total rate is signi cantly large and more dominant in the relevant range of transverse momentum. hus the value above $\gamma_q >$ γ_g, uark ow parameters is not suitably t to get the photon spectra. At the same time, it is also not right to take low value of uark phenomenological parameter i.e. $\gamma_q < 2\gamma_g$ since for these values the calculation for photon rate diverges. e compare our work with other theoretical work at ero chemical potential 4 , . herefore, the results are much enhanced in the presence of chemical potential as comparison to ero chemical potential of other works 4 , .

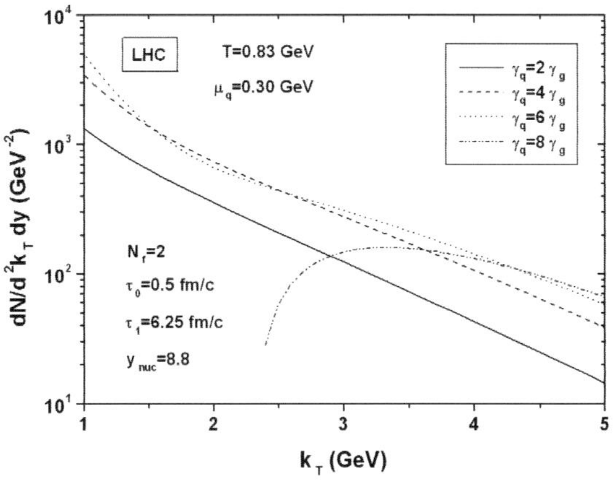

Figure 2. Photon Spectra at temperature T . 3 GeV using various value of uark ow parameter dependent on uark chemical potential at .

In Figure 2 , with the similar line, we plot total photon spectra for at hot temperature T . 3 GeV and at value of uark chemical potential μ_q .3 GeV for avor N_f 2. It is seen that the photon rate at is much higher than the R I data. At , the enhacement in the photon yield is of the order of 1 ^{1} to 1 ^{2}. So, the production rate at is signi cantly large as comparison to R I energy.

Again we observe that the production rate is low for the value of ow parameter γ_q γ_g up to transverse momentum k_T 2. GeV and for γ_q γ_g up to k_T 4. GeV and afterwards photon rate increases. he total production rate is very large and more dominant in the relevant range of transverse momentum. Overall our model results are suitably t for γ_q $2\gamma_g$ to γ_q γ_g with various sets of parameter. In this case, the results are much enhanced in the presence of chemical potential at as comparison to R I energy and production rate of photon at non-vanishing chemical potential is much larger than the ero chemical potential.

Finally, we conclude that the total photon yield is calculated by integrating the rate over the plasma volume created by the e pansion and is compared with other theoretical work 4 , at ero chemical potential. he measurement of leading order processes for photon production in the GP which consist D ompton Scattering, uark-anti uark annihilation, bremsstrahlung, and uark-anti uark annihilation with scattering provide a good opportunity to study the photon radiation from hot and dense matter of GP.

At both R I and energies, total photon yield is appreciably large with the e ect of uark phenomenological ow parameter and hence it is a strong increasing function at high temperature and chemical potential. So, our results are tted well with these sets of parameter and show appreciable enhancement in the production of photon. y considering the uark ow parameter in uark mass, photon measurements may play an important role.

Overall photons are the best probe of GP in relativistic heavy-ion collisions.

3.1. Acknowledgments

he author acknowledges the conference organi ers for the opportunity to present this talk and for a most stimulating conference. P Jain is very grateful to the department for giving the opportunities to do research.

References

1 Arsene I *et al.* 2 *Nucl. Phys. A* **757** 1.
2 Andronic A, raun- un inger P and Stachel 2 *Nucl. Phys. A* **772** 1 .
3 Ste an F D and homa 2 1 *Phys. Lett. B* **510** ra ler and homa 1 *Phys. Rev. C* **53** 134 .
4 clerran and oimela 1 *Phys. Rev. D* **31** 4 .
Peit mann and homa 2 2 *Phys. Rep.* **364** 1 .
Steele V, amagishi and ahed I 1 *Phys. Lett. B* **384** Steele V and ahed I 1 *Phys. Rev. D* **60** 3 2.
Dusling , eaney D and ahed I 2 *Phys. Rev. C* **75** 24 Dusling k and ahed I 2 1 *Phys. Rev. C* **82** 4 .
Arnold P , oore G D, and a e G 2 2 *JHEP* **0206** 3 Arnold P, oore G D and a e G 2 1 *JHEP* **0111** .
hatter ee R, Frodermann S, ein and Srivastava D 2 *Phys. Rev. Lett.* **96** 2 23 2.
1 ratkovskaya , iselev S and Sharkov G 2 *Phys. Rev. C* **78** 34 .
11 ees V, Gale and Rapp R 2 11 *Phys. Rev. C* **84** 4 .
12 asar G, har eev D and Skokov V 2 12 *Phys. Rev. Lett.* **109** 2 23 3.
13 iu F and iu S 2 14 *Phys. Rev. C* **89** 34 .
14 Shen , ein , Pa uet F and Gale 2 14 *Phys. Rev. C* **89** 44 1 .
1 innyk O, onchakovski V, assing and ratkovskaya 2 13 *Phys. Rev. C* **88** 34 4 innyk O, assing and ratkovskaya 2 14 *Phys. Rev. C* **89** 34 .
1 Dion , Pa uet F, Schenke , oung , eon S *et al.* 2 11 *Phys. Rev. C* **84** 4 1.
1 lasen , lein- oesing , nig F and essels 2 13 *JHEP* **1310** 11 .
1 Adare A *et al.* P NI ollaboration 2 12 *Phys. Rev. Lett.* **109** 1223 2.
1 homa 2 *Phys. Rev. D* **72** 4 3 .
2 iu F , irano , erner and hu 2 *Nucl. Phys. A* **830** .
21 Dion *et al.* 2 11 *Phys. Rev. C* **84** 4 1.
22 hatter ee R *et al.* 2 13 *Phys. Rev. C* **88** 34 1.
23 Singh S S and umar 2 14 *Int. J. Mod. Phys. A* **29** 14 11 .
24 apusta , ichard P and Seibert D 1 1 *Phys. Rev. D* **44** 2 4 apusta , ichard P and Seibert D 1 3 *Phys. Rev. D* **47** 41 1 apusta and clerran D 1 2 *Phys. Lett. B* **283** 14 .
2 aier R, Nakkagawa , Neigawa A and Redlich 1 2 *Z. Phys. C* **53** 433 Neubert 1 *Z. Phys. C* **42** 231.
2 Sinha 1 *Phys. Lett. B* **157** 221 Alam , Sarkar S, atsuda , Nayak and Sinha 2 1 *Phys. Rev. C* **63** 21 1 R .
2 Gale and apusta I 1 1 *Nucl. Phys. B* **357** .
2 orken D 1 3 *Phys. Rev. D* **27** 14 .
2 Nagamiya S 1 2 *Nucl. Phys. A* **544** c.
3 Dumitru A *et al.* 1 3 *Mod. Phys. Lett. A* **8** 12 1.
31 ammon N *et al.* 1 *Phys. Rev. C* **61** 14 1.
32 Geiger and apusta I 1 3 *Phys. Rev. D* **47** 4 .
33 ra ler , Vi a and homa 1 *Phys. Lett. B* **346** 32 .
34 Arbe N, Ornik , Plumer , immermann A and einer R 1 *Phys. Lett. B* **345** 3 .
3 Dutta D, ohanty A , umar and houdhury R 2 *Phys. Rev. C* **61** 4 11 Dutta D, Sastry S V S, ohanty A , umar and houdhury R 2 3 *Pram. J. Phys.* **60** 1 .
3 e , ong , a , u and iu 2 *Phys. Lett. B* **628** 2 .
3 e , ong , a and a G 2 *C. Phys. Lett.* **22** 13 .
3 ong , e , a and iu 2 *Nucl. Phys. A* **766** 2 1
3 ong , e , a and iu 2 *Phys. Rev. C* **72** 4 .
4 ong , e and a 2 *C. Phys. Lett.* **23** .
41 Gervais and eon S 2 12 *Phys. Rev. C* **86** 34 4.
42 Singh S S and umar 2 1 *Int. J. Mod. Phys. A* **30** 1 2 .
43 iro S, Doorn V, uller , homas and ang N 1 3 *Phys. Rev. C* **48** 12 .
44 Strickland 1 4 *Phys. Lett. B* **331** 24 .
4 umar and Singh S S 2 13 *ISRN High E. Phys.* **2013** 1 4 .
4 umar and Singh S S 2 12 *Can. J. Phys.* **90** .
4 Gosain D S, Singh S S and ha A 2 12 *Pram. J. Phys.* **78** 1 .
4 Aurenche P, Gelis F and araket 2 *Phys. Rev. D* **61** 11 1 Aurenche P, Gelis F and araket 2 *Phys. Rev. D* **62** 12.

4 Arnold P, oore G D and a e G 2 1 *JHEP* **0112** .
Renk 2 3 *Phys. Rev. C* **67** 4 1.

Institute of Physics
Dirac House, Temple Back
Bristol BS1 6BE UK

ISSN: 1742-6588
ISBN 978-1-5108-1195-9

CURRAN ASSOCIATES INC.
proceedings
.com

9 781510 811959